The
GREAT
NERVE

The
GREAT
NERVE

*The New Science of the Vagus Nerve
and How to Harness Its Healing Reflexes*

DR KEVIN J. TRACEY

PENGUIN LIFE

AN IMPRINT OF

PENGUIN BOOKS

PENGUIN LIFE

UK | USA | Canada | Ireland | Australia
India | New Zealand | South Africa

Penguin Life is part of the Penguin Random House group of companies whose addresses can be found at global.penguinrandomhouse.com.

Penguin Random House UK,
One Embassy Gardens, 8 Viaduct Gardens, London SW11 7BW

penguin.co.uk

First published in the United States by Avery,
an imprint of Penguin Random House LLC 2025
Published in Great Britain by Penguin Life 2025
001

Copyright © Osprey Holdings, Inc, 2025

The moral right of the author has been asserted

Penguin Random House values and supports copyright. Copyright fuels creativity, encourages diverse voices, promotes freedom of expression and supports a vibrant culture. Thank you for purchasing an authorized edition of this book and for respecting intellectual property laws by not reproducing, scanning or distributing any part of it by any means without permission. You are supporting authors and enabling Penguin Random House to continue to publish books for everyone. No part of this book may be used or reproduced in any manner for the purpose of training artificial intelligence technologies or systems. In accordance with Article 4(3) of the DSM Directive 2019/790, Penguin Random House expressly reserves this work from the text and data mining exception

Printed and bound in Great Britain by Clays Ltd, Elcograf S.p.A.

The authorized representative in the EEA is Penguin Random House Ireland, Morrison Chambers, 32 Nassau Street, Dublin D02 YH68

A CIP catalogue record for this book is available from the British Library

ISBN: 978-0-241-76240-0

Penguin Random House is committed to a sustainable future for our business, our readers and our planet. This book is made from Forest Stewardship Council® certified paper.

To Grandpa Culotta

CONTENTS

Introduction: A New Frontier in Medicine   ix

I
GREAT SECRETS:
Understanding the Hidden Power of the Vagus Nerve

1. How Electricity Could Replace Your Medications   3
2. The Great Nerve Reveals Itself   21
3. Your Body's Healing Reflexes   43

II
GREAT INTERVENTIONS:
The New Frontier of Stimulation Therapy

4. The Path to Stimulation and Early Experiments with Epilepsy   73
5. Rebalancing Inflammation   93
6. Beyond Medication: A Healing Reflex for Depression   115
7. Outside-In Stimulation to Regulate Body Weight, Treat Diabetes, and More   137
8. The Ear-Brain-Body Connection: Over-the-Counter Devices with Many Potential Benefits   155

III

GREAT EXPECTATIONS:
Everyday Tools with Promise

9. Meditation and Breathwork	*179*
10. Cold and Exercise	*205*
11. Your Great Nerve: How to Talk to Your Doctor (the FAQs)	*227*
Coda: The Clear and Present Future: Computer Chips, Not Medicines	*247*
Acknowledgments	*259*
Appendix: A Bioelectronic Snapshot: The Current Status of Vagus Nerve Stimulation and Related Therapies for Various Conditions	*263*
Notes	*267*
Index	*295*

Introduction

A New Frontier in Medicine

On November 15, 2011, I met the first patient treated with something I had invented. When he was still a young man, an autoimmune disease had sentenced him to disability and premature death. Rendered homebound by debilitating pain in his arms and legs, he could not work or even play with his young children. But in 2011, a neurosurgeon implanted him with a pacemaker-like device under his left collarbone. The surgeon tunneled under the skin to thread a wire up into his neck, connecting the pacemaker to his vagus nerve. The patient's pain, inflammation, and disability disappeared after he received his vagus nerve stimulator.

This new field of vagus nerve stimulation therapy to treat inflammation is now poised to revolutionize the way millions of other people are cared for. Its widespread use can help prevent inflammation that contributes to the mortality of forty million people worldwide every year. Even when it is not fatal, inflammation impairs the health of millions more. Inflammation has replaced infection as the greatest threat to healthful human longevity. I wrote this book to give you the essential information you need about your vagus nerve, including how it regulates inflammation in your body and how you can help it do its job, from everyday healthy habits to the latest medical technology.

Here you will find a new way to think about your immune system. Traditionally, it was studied on its own as an independently functioning system that polices itself and does its own bidding to ward off threats from infection and injury. Scientists and physicians viewed the immune system outside the domain and control of the brain and nervous system, though the latter were understood to coordinate the functions of all your other organ systems. But a discovery in my lab revealed that the brain and immune system are inextricably linked through the vagus nerve. This was the critical insight that enabled my colleagues and me to invent the vagus nerve stimulating device that gave the first patient a second chance at a full life in 2011.

The vagus nerve has been studied for more than two thousand years, intriguing scientists with its great reach, but it continues to hold many secrets, which hundreds of laboratories around the world are working to unravel. With each passing week, new information emerges, providing evidence that points to new ways of tapping into the vagus nerve's ability to regulate our vital systems and heal us. Our discoveries are guiding us to develop new therapies for inflammation, depression, anxiety, epilepsy, substance use disorders, headaches, cardiovascular and gastrointestinal diseases, Alzheimer's disease, Parkinson's disease, stroke, multiple sclerosis, and other conditions. There are thousands of scientific and medical publications addressing these topics, and my colleagues and I have written many peer-reviewed research papers and review articles ourselves. You'll find a sample of these cited in the endnotes for those who want to dive into detailed discussions of molecular mechanics. The purpose of this book is not to recap all this detail but to introduce you to your vagus nerve and give you enough information to begin tapping into its greatness yourself.

Often described today as a "superhighway" between the brain and the body, the vagus nerve turns out to be much more mysterious and beautiful than traffic. Like the strings, reeds, skins, and soundboards of an orchestra, the two hundred thousand or so nerve fibers of your vagus vibrate in tune with health. The vibrations are the music, and the song is life.

If you are fairly healthy but overwhelmed by the billions of web impressions, #vagusnerve, and claims on the internet and social media telling you to do this, that, or the other thing to stimulate your vagus nerve, I understand. I also find the advice swirling "out there" to be a lot. Worse, it is often inaccurate and sometimes completely wrong. Many popular recommendations and proposed therapies, if they do stimulate your vagus nerve at all, may or may not have proven health benefits. But some of these ideas are grounded in verifiable scientific mechanisms and physiological understanding, warranting additional thought and discussion. Information in this book will help you separate the facts from the hype. As a neurosurgeon-scientist, I do my best to break down complex concepts into easy-to-understand terms to explain what it is we do know—as well as what we don't know, yet.

If you or someone you care about is, unfortunately, unwell and perhaps suffering from a condition caused by inflammation, then my sincere hope is that this book will give you a basic understanding of the role of your (or their) vagus nerve in that condition, and at least enough insight so that you feel empowered to

INTRODUCTION

ask the right questions of your healthcare providers. We are in the early decades of a revolution in medicine, one that is launching previously unimaginable computerized electronic therapies to treat dangerous and disabling illnesses. I wrote this book to tell you about them.

The time for this book is now. As I write, large clinical trials are completed or nearing completion, and vagus nerve stimulation may soon become widely available as a therapeutic tool to treat inflammatory conditions. There is more work to do, more knowledge to know, and more clinical trial results to be collected. But an avalanche of promising new data is barreling down the mountain on a path that I expect will push the adoption of vagus nerve stimulation therapy into general medical practice. I also believe this breakthrough will benefit millions of people.

I have had the privilege and great fortune to be an institution founder and director in both the nonprofit and for-profit sectors, including SetPoint Medical, a global leader in the bioelectronics and vagus nerve stimulation industry. Since 2005, I have served as the president and CEO of the Feinstein Institutes for Medical Research at Northwell Health in New York, the home of more than one hundred principal investigators and labs, staffed and supported by 8,500 team members pursuing the Institutes' mission of "producing knowledge to cure disease." Feinstein is a destination for research and clinical translation and is recognized internationally for leadership in bioelectronic medicine, immunology, behavioral health, health system sciences, and other areas of science and medicine. It is a collaborative and welcoming home for creative scientists and physicians from all over the world. Together these researchers, scientists, physicians, postdocs, students, and staff are driven to produce knowledge not simply for the sake of knowledge but to make discoveries that pave the way to innovative new cures and treatments to improve human health and create a better future for all of us.

Full disclosure: Although my original patents for the idea of using vagus nerve stimulation to treat inflammation led me to cofound a new company to test this idea in patients, I do not have an insider role in SetPoint Medical. But I do continue to serve as a consultant to the company in the hope that the vagus nerve stimulator they developed, based on my original patents, becomes widely available for those many patients in need. Back in my lab, my colleagues and I continue to make new inventions in the field of bioelectronic medicine to treat inflammation, and my rights for these inventions are assigned to my employer, the nonprofit Feinstein Institutes.

In my forty-year fascination with the brain and inflammation, I have learned that the vagus nerve holds the potential to alleviate pain and suffering. This is why I go to work in the lab every day, where I collaborate with some of the brightest and most innovative, enthusiastic, and persistent individuals on the planet, driven by our shared mission to produce knowledge that is useful.

This book brings together all that we now know about the vagus nerve and the revolutionary potential that it holds for our health. I have divided it into three parts. In part 1, Great Secrets, you'll meet the vagus nerve in all its humming, sprawling glory through stories of its structure and functions. If your primary interest is in practical applications that may help you or someone you love, feel free to skip ahead to part 3 (and circle back to part 1 later). Part 1 will chart the history of our understanding of the vagus nerve and reveal our modern knowledge of its inner workings using modern tools in microneurosurgery, optogenetics, molecular biology, and neural decoding with artificial intelligence. Having a functioning vagus nerve is truly a life-or-death matter, and not only because it's the only nerve we have that when it is cut (on both sides), we die. It's responsible for balancing the day-to-day and cell-by-cell functioning of our vital organs and vital systems, including our immune system. It does all this through its reflexes.

Part 2, Great Interventions, will take you to the labs and lives of people applying what we now know about the vagus nerve to treat diseases in ways we never thought possible. In some cases, I have changed the names and identifying details to protect privacy, and I have included citations for the work of my colleagues and historical work for those who seek more detail. As this book goes to press, a revolutionary breakthrough in bioelectronic therapies is nearing completion, which shrinks the device to the size of a Tylenol capsule (including its rechargeable battery) that can be implanted onto the vagus nerve in the neck to stimulate an anti-inflammatory healing reflex and check inflammation. I'll describe how we established the exciting new frontier of bioelectronic medicine, where we are today, and why we already know it's safe. Vagus nerve stimulators have been around and in use by more people for longer than you probably imagined: Since the 1990s, hundreds of thousands of people have received FDA-approved vagus nerve implants to treat epilepsy and depression.

More recently, clinical trials of vagus nerve reflexes that govern appetite, metabolism, insulin, and glucose control are studying whether noninvasive focused ultrasound can be used to treat diabetes and obesity. And there are hundreds of clinical trials globally exploring indirect pathways to the vagus nerve that do not require surgery but instead aim to "hack" the brain (gently) with devices that sit on the ear to stimulate a branch of the vagus nerve there, potentially treating depression, anxiety, migraines, headaches, opioid withdrawal, sleep disorders, and COVID-related challenges, as well as enhancing memory and improving cognitive function. At the forefront of this exciting exploration, my team at the Feinstein Institutes is partnering with industry giants and research collaborators at dozens of other institutions.

Part 3, Great Expectations, turns a scientific eye on popular at-home, DIY practices such as meditation, cold plunges, and exercise. As a neurosurgeon-scientist I am concerned about making the proper diagnosis, answering complex questions based on the best available data, and providing the safest advice for patient benefit with minimal risk. And so in the last part of the book, I am careful to offer practical tools that are promising as the science continues to emerge. From ancient practices to the latest trends on TikTok, people have long experimented with mind-body interventions and homemade ways to tap into their vagus nerves. With the Hippocratic oath, available evidence, and curiosity as my guides, I consider advice from the Dalai Lama, the Dutch endurance athlete Wim Hof, and other experts, and use common sense to untangle the vagus nerve science from the hype when it comes to meditation, breath control, ice baths, exercise, and over-the-counter devices. I will explain what I find true and useful, what is false or even dangerous, and what we just don't know yet.

As science and innovation converge on the vagus nerve in medicine and everyday life, we are coming to understand not only how to maintain it but also why our health and well-being depend on it. I hope to empower you with knowledge and inspire you with true stories of people and research in progress. The stories in this book are true, and they are drawn from my personal experiences as well as from those who shared their stories directly with me, for which I am deeply grateful. In some cases, I have changed the names and identifying details to protect privacy, and I have included citations to the work of my colleagues and to historical work for those who seek more detail.

Empowered with knowledge, you can take charge of your health and make informed decisions about your well-being. The first step to taking care of your vagus nerve is knowing you have one. The next step is understanding what it does, what you can do to keep it healthy (so it can keep you healthy), and what bioelectronic medicine can do for you. Then you can ask questions and make choices to take care of your vagus nerve on a daily basis and, if you have a serious medical condition, to find out about options for treatment. Learning about the great nerve inches us closer to a world in which catastrophic inflammation can be averted and inflammatory diseases that affect millions are better treated.

I
Great Secrets

Understanding the Hidden Power of the Vagus Nerve

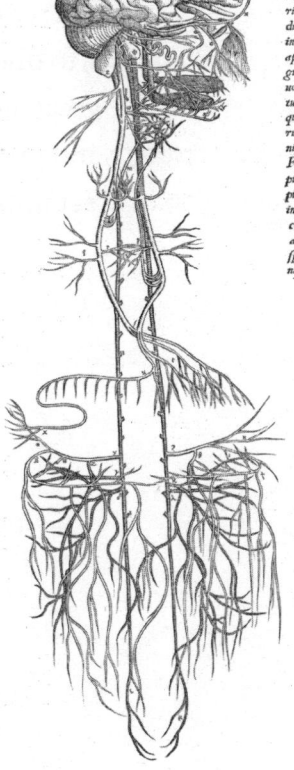

I

How Electricity Could Replace Your Medications

Suffering has been stronger than all other teaching.
—CHARLES DICKENS, *GREAT EXPECTATIONS*

The email caught my eye, subject line: "Thank you for saving my life." It was written by a woman named Kelly Owens, and it told her story. When she was thirteen, Kelly was diagnosed with Crohn's disease and inflammatory arthritis, the beginning of fifteen years of illness, hospitalizations, surgeries, immobility, and immune-suppressing medications. None of them helped her. By her late twenties, she had been treated with steroids for so long that she developed osteoporosis. She was a young woman with the bone density of someone three times older, a newlywed whose young husband had gotten used to hoisting her on his back to carry her from a restaurant on occasions when her legs swelled so much during dinner that she couldn't walk to the car.

In 2014, Kelly saw me on HuffPost Live and soon afterward

emailed me for the first time. I replied with information about our work on the vagus nerve and suggested that she keep track of the clinical trials SetPoint Medical was planning. Three years later, Kelly's doctors told her that they were out of options and, worse, that she should go home and prepare for a life completely dependent on the steroid prednisone.

The right amount of inflammation is the delicate fulcrum between your body's frontline defenses against infection and your own immune system wreaking havoc on the body it is meant to protect. Kelly's immune system had lost its balance, with catastrophic effects. Despite relying on prednisone to get through each day, her legs were so swollen she often couldn't walk, forcing her to give up a teaching career. The frequent flare-ups of her Crohn's disease meant she missed too much work. As her last resort, crying, she searched clinicaltrials.gov, where she learned that SetPoint was performing clinical trials in Europe. Paying for the medications that had failed to help had wrecked their finances, but with the support of friends and family, Kelly and her husband, Sean, raised enough money to move. Then they packed up their life and relocated overseas, moving from New Jersey to Amsterdam, the Netherlands.

At Amsterdam University Medical Center, Kelly received a small electronic device implanted beneath her collarbone that would deliver electrical signals to stimulate her left vagus nerve and, we hoped, activate her anti-inflammatory reflex to stop her excessive inflammation. Within weeks, Kelly was walking without a cane for the first time since she was a teenager. Eight weeks later, she was in clinical remission, free of gastrointestinal symptoms, running two miles a day, and working out. A few months later she was finally weaned off prednisone, leaving her medication-free.

I still hadn't met Kelly at that point, but tears blurred the words on the screen as I read her email and learned of her triumph. Imagine living through her suffering—or maybe you don't have to imagine. Maybe you, like Kelly, are one of the more than five hundred million people worldwide with an autoimmune disease. Or maybe like Kelly's husband and father and mother, you love someone who is. Maybe you don't have a diagnosis, but inflammation affects you in some other way. In fact, inflammation affects nearly all of us.

Kelly's email update continued: "Now that we have returned, I am planning the rest of my life—something I haven't been able to do in years."

I have been fortunate to befriend Kelly in the years since, and we even became colleagues when she made it her mission to advocate for access to bioelectronic medicine for everyone who might benefit. She gave me the cane she doesn't need anymore as a token of her gratitude for the medical research that led to the small implant next to her collarbone, safe as a pacemaker, and gave her back her life. The cane leans against the bookshelf in my office with the same big white bow attached to the handle as when she presented it to me, a card from her still taped to the side. The shaft of the cane is swirly hot pink, and every time I look at it, I think how perfect and how terrible to get a cheerful pink cane to help you walk when you're a teenager.

Six years after Kelly's grateful email, SetPoint Medical continues clinical testing of a newer, even tinier vagus nerve stimulating device to treat people with rheumatoid arthritis in the United States. The Food and Drug Administration (FDA) has assigned a breakthrough designation to the technology, and my hope is that it will soon enter the treatment armamentarium for physicians to prescribe nationwide. Patients are clamoring for a new option to treat inflammation. Their current therapies are expensive, are often administered as injections, have dangerous side effects, and don't always help. The side effects of these powerful immunosuppressant drugs are dangerous enough to warrant a "black box warning" from the FDA, the direst red flag available when one possible side effect is death. The advent of vagus nerve stimulators, like Kelly's, raises hope for future patients being offered safer alternatives to today's limited treatments.

Today, thanks to antibiotics, vaccines, and modern sanitation, we reasonably expect to live to old age. Most of us die from noninfectious causes: heart disease and stroke, diabetes and obesity, neurodegeneration as in Alzheimer's disease and Parkinson's, and cancer account for *two-thirds* of the sixty million human deaths each year. That statistic may not have surprised you, but what many people don't know is that these are all diseases of inflammation. And this tally does not count the millions more who suffer from autoimmune diseases and other inflammatory conditions during their lives.

What if we could influence the vagus nerve's own signals to help it do what Nature designed it to do? Can stimulating the vagus nerve help people with seemingly untreatable diseases? Can it help those of us who want to keep our bodies and minds working and thriving better and longer? What if there were things you could do to keep your vagus nerve in tune? What if you could communicate with your immune system or regulate your insulin and glucose with a computer chip? When my colleagues and I discovered, a few decades ago, that the harmonious cooperation of your immune and other vital systems depends on communication between the brain and the body through the vagus nerve, we began exploring ways to participate in that kind of music. We have found many—with measurable results that I will tell you about.

The short answer is: We can play the vagus nerve as nature intended, with electricity, sound waves, and computer chips, to save lives like Kelly's.

SOMEBODY SHOULD DO SOMETHING ABOUT THAT

When she was twenty-nine years old and I was five, my mother, Dorothy, died from a brain tumor on a cloudy summer Sunday. Because my father could not bring himself to tell me, I did not know it until the following Friday, after the funeral. My brother, Tim, was four; my sister, Sharon, was not yet two, too young to understand. Tim and I learned our new fortune sitting in Dad's lap in the living room of the little cape-style house where he had lived until he married Mom. I don't know how long we sobbed in his arms, but when we stopped, he handed us each a new toy Matchbox car. Clutching mine, I ran from the gloom into the sunny backyard.

A week or so earlier, Mom took a long afternoon nap. Bored, Tim and I began playing "telephone line repairmen" in the house, hammering nails and stringing string directly into the wood paneling that lined the staircase (a rare luxury for my just-starting-out parents). When our father arrived home from work and saw us, he yelled loud enough to shatter our fantasy, but Mom continued sleeping through it all.

Now, I understand. A *glioblastoma multiforme* was busily compressing her brain and cutting off her nervous system. She had been fighting, unbeknownst to any of us, cancer. Headaches for months, uncharacteristic lapses of memory, the steady decay of her once graceful handwriting, a tumor was encroaching on her very existence, constricting and distorting the signals between her brain and body.

I have only a few blurry dark memories from what happened next. We were crammed into a Plymouth station wagon on a steamy, hot five-hour ride to Hamden, Connecticut. I complained about the heat and the absence of air conditioning. My father drove and told me to be quiet. In the passenger seat next to him, my mother just slept.

I was riding behind her. Tim sat behind my father, and little Sharon lay between us in her bassinet, which wobbled and threatened to capsize with every bump in the road. (It was the 1960s.) I remember watching the back of my mother's head for hours. Her black hair fell across her shoulders and grazed the back of her seat, just out of the reach of my fingertips. That is my last memory of her, and it sounds dramatic because it was. I never saw her again.

My father had hustled us all into the car at the urging of a doctor who had examined my mother. We were driving to see my grandfather Culotta, my mother's father, who was a professor of medicine at Yale.

Now, I understand. My mother had severe papilledema, the condition that occurs when pressure inside the brain becomes dangerously high. This pressure was pushing through her optic nerve from her brain into the back of her eye. Using an ophthalmoscope, with its built-in light and magnifying lens, her local doctor had seen her retina and found the spot where the optic nerve is attached. This spot is called the *optic disc*, and it normally looks like a bright yellow full moon, with a crisp edge, shining in a red retinal sky.

But looking into my mother's eyes, the doctor saw a misshapen blob, with a blurry perimeter shrouded in fog. He recognized that pressure and swelling in my mother's brain were forcing fluid into her optic disc, and into all the other nerves connecting her brain to her body. At Yale, Dr. William J. German, the chief of neurosurgery, operated. Suspecting a tumor, he performed a craniotomy, removing a four-by-four-inch circle of bone from her skull.

To a neurosurgeon, the normal human brain is a captivating masterpiece: supple texture, subtly pink, elegant, and graceful. A Degas, perhaps. The hole in the skull is a window to an intricate web of blood vessels that nourishes and sustains billions of neurons suspended in glistening soft tissue. To me, none of this is "gross" because in the glow of an operating theater, there's a strange combination of wonder and the feeling that it's time to get down to business. I have a job to do, and at the same time, I'm aware that each nerve cell arrayed before me has its own dominion, a unique portal into an overlapping network of signals that produce your memories, thoughts, emotions, feelings, actions, and reflexes—your every breath and every heartbeat—all of it.

My mother's brain was a different picture. Tumors are gross. Requiring proof of the diagnosis, Dr. German would have excised a small piece of her tumor and handed it off to the pathology lab to be studied under the microscope. Then, as I have done hundreds of times in my own neurosurgery operating room, he stood by her head, waiting for the phone to ring. I am sure he already knew the prognosis. He knew that later in the waiting room he would deliver the worst news to my father and grandfather. And as neurosurgeons still do today, he hoped he was wrong.

In 1963, as in 2023, a diagnosis of glioblastoma multiforme is terrible news. Survival is measured in weeks and months, not years. There was and is no effective therapy to stop the growth that pressurizes the brain, trapped as it is in the skull. Even now, sixty years later, the median survival time is twelve to fifteen months. So, in Dorothy's operating room, when the phone rang, there was nothing else for Dr. German to do except close up the opening he had made. He wrapped her head in bandages and moved her to the recovery room, but my mother never woke up again. Her brain continued to swell, and she died that Sunday.

A few days after the funeral, I remember climbing up into Grandpa's lap. Breathing the sweet cigar scent from his shirt and the green leather of his wingback chair where he read medical journals, I asked why his friend Dr. German hadn't fixed my mother and sent her home.

"The tumor growing in her brain was large, and it had lots of legs," he said, "shooting off in many directions. There was no way he could pull out all those legs without damaging the normal parts of her brain. She'd be paralyzed and unable to speak. Your mom wouldn't have been who she was. There was nothing anyone could have done."

"So," I said, "Somebody should do something about that."

Nodding, he whispered, "Well, maybe someday you will."

This turned out to be good career advice.

INPUTS AND OUTPUTS

Although there are negative consequences from losing a parent during your childhood, other outcomes are also possible. A five-year-old learns that life doesn't last forever. It can end at any time.

So, faced with this certainty, we may ask ourselves two fundamental questions about the rest of our lives. What is it we want to do? And how will we use the time we have on Earth? My five-year-old self decided that if I could, I wanted to help other kids not lose their mothers. As I grew up and became a neurosurgeon, however, the scope of my interests expanded from mothers to, simply, others.

Some moments in medical school are truly surprising. After years of study, you have a firm foundation for what you think you know until a new fact or idea upends your world. This can happen when a teacher reveals a surprising new insight or when you witness something for the first time. The human body provides unlimited mysteries, some of which can hit you with stunning force. I remember first seeing how organs in the human body are organized into compartments, for example, how the separate fasciae of your chest and abdomen function like dividers in a suitcase, keeping everything neatly in its place. Or the way, on another day, our anatomy professor, Dr. William McNally, had reached his gloveless hands into the chest of my cadaver and, up to his elbows, separated the lobes of a smoker's black lung with his own yellow smoker's fingers to show us what chronic obstructive pulmonary disease and emphysema look like, all of us aware that his lungs probably looked like that too.

Most of all I remember the time Dr. McNally described the human nervous system, which is the way I describe it to this day. To understand why the vagus nerve matters, not only to Kelly and others looking for relief from disease but to every one of us living in a human body and hoping for a long, healthful, and happy life, you need to know how your whole body depends on this nerve. I stress *whole body* because the brain is so often pulled into focus during discussions of health and well-being, making it too easy to forget that the brain is an inseparable *part* of the body, reaching as you will see into your most vital systems via your vagus nerve.

To understand why your vagus nerve matters, you need to know that your nervous system works in two directions: Your sensory system provides the input (from the brain's perspective), and your motor system provides the output (from the brain to the body). Sensory nerves transmit information from the body to the brain; motor nerves send information from the brain back to the body. So far, so good.

But on the motor side, there are two kinds of outputs. There's the stuff you can control, like when you sign your name on the dotted line, shoot a basketball, or squeeze the hand of someone you love, and the stuff you can't, like when your heart beats and how fast. So, consciously and unconsciously, motor nerves transmit information from your brain to your body, whereas sensory nerves gather information about what's happening inside and outside your body and send it back to your brain.

Input and output, what you can and can't control; and the brain and body communicating with each other, constantly, each making adjustments depending on what it "hears" from the other, so much of this happening automatically. Your health is a product—an emergent property—of your sensory and motor nerves working together to harmonize your existence in a perpetual exchange between the brain and body, each adapting and recalibrating based on the other's signals. It's astonishing, when you think about it. So much of this sophisticated interplay happens in the background, *without* your having to think about it, unnoticed in your daily life.

The part of the nervous system that you don't have to think

about is called the *autonomic* nervous system, from the same root as *autonomy*, as in self-government. The autonomic system coordinates your heart, lungs, and intestines, your pancreas and liver and salivary glands, and all the other organs that work around the clock to keep you alive and in physiological balance. And this part of the nervous system that we don't have to think about itself has two sides, which will sound familiar whether or not you have any medical training: "fight-or-flight," as we say about the sympathetic nervous system responsible for the hyper-readiness, speed, and strength that we need in case of danger, and "rest-and-digest" to describe the parasympathetic system that keeps us calm, cool, and collected.

Resting, digesting, and keeping calm all depend on your vagus nerve, which is the linchpin of the parasympathetic system. Professor McNally's class on the nervous system was the first I'd heard of this nerve. More than forty-five years after I heard that strange word *vagus* for the first time, I can tell you that what most people talk about when they talk about the vagus nerve is not what's most interesting, or based on the latest research, or what really matters for your health.

The medical textbooks and common dogma have some catching up to do, even though some of my lab's most groundbreaking discoveries are more than twenty-five years old. They'll tell you the vagus nerve is known as the tenth cranial nerve or cranial nerve X; that it's the longest nerve in your body; that its name is assumed to derive from the Latin term *vagus*, meaning "wandering," a nod to its extensive reach from your brain stem at the base of your brain, where it exits your skull at about the level of your ears and runs down either side of your neck before branching on out to the organs throughout your chest and abdomen. You'll learn that although we refer to it in the singular, your vagus nerve is actually two nerves, one on each side, what anatomists call a "paired structure," just as you have two kidneys, two legs, and two thumbs. You'll learn that, like the rest of the nervous system, the vagus goes both ways because it is really a bundle of nerve fibers (two hundred thousand total, or one hundred thousand on each side), both sensory and motor, input and output, forming a system of two-way communication between the body and the brain as they coordinate so many of your vital functions.

All these facts are true, but are they really why the vagus nerve matters? Before it was called the *vagus*, for many centuries, this nerve was simply called *great*, as in "the great nerve"—*your* great nerve. The question always has been, *how* great? Since the textbooks don't provide all the answers, my colleagues and I dedicate ourselves to unraveling its secrets in the lab each day.

Kelly's cane that she no longer needs reminds me, in the middle of a busy day, of those questions that first struck me as a boy who lost his mother. Medical research that leads to therapies and cures is what I want to do. This is how I choose to spend my time. From my grandfather to my mother to me and everyone I've worked with who stayed curious and asked questions beyond the textbooks, to Kelly and millions of others who are suffering, I like to think we pass a baton of knowledge and compassion, kind of like Kelly passing her cane to me.

EUREKA: DISCOVERING THE TWO-WAY REFLEX

There's an old medical school joke that professors used to tell. When immunology was the next topic in the curriculum, they would say the neuroscience students could have the day off, and vice versa for immunology students during neuroscience. The joke worked because the nervous system and the immune system were thought to be firewalled: separate, no communication between them—white blood cells floating around in their own space having nothing to do with neurons, and neurons going about the brain's business, which certainly didn't include immunity and inflammation. This separation was the prevailing understanding in the 1980s and '90s. However, my research for more than four decades has overturned this conventional wisdom, revealing not only communication but lifesaving implications that have made the joke irrelevant.

It all started with the biggest surprise I have had in a lab. We were studying the effects of brain damage in rats. To do this, we administered an experimental drug that we had designed to prevent brain inflammation, but we found that putting the drug in the brain affected the immune system and stopped inflammation throughout the body, and that the vagus nerve was the reason. The idea was revolutionary, yet it just made sense when you understand that communication in the nervous system goes both ways. The body, vagus nerve, and brain not only react to injury and infection (input) but respond with signals that regulate the immune system's inflammatory response (output)! Using advanced technologies from molecular biology and Silicon Valley, we went on to prove this to be true. My colleagues and I discovered the motor arc of what turns out to be a two-way reflex, including a neural output that allows us to *work with the vagus nerve to control the immune system*. It is a healing reflex. That we can work with.

When you are healthy, that part of the nervous system you don't have to think about self-governs. When it doesn't, you become ill. An intricate network of nerves is essential for maintaining balance and harmony within the body, and understanding how the vagus nerve normally keeps you healthy is a new key to understanding how to intervene to treat disease. In other words, the vagus nerve knows what it's doing to keep your body in harmony. By understanding how it does this, we can intervene to fix things if they go wrong.

When we published our discovery in *Nature*, we described for the first time how the nervous system reflexively tunes the inflammatory response in real time but also suggested an opportunity to retune the vagus nerve when our immune systems go out of tune. This discovery, coupled with ongoing research and technological innovation, ignited the new field of bioelectronic medicine, which is unraveling new mechanisms to work with the nervous system using electronic devices to alleviate suffering and stimulate healing. The ability to manipulate and prevent specific disease processes by using electrical impulses to stimulate the vagus nerve makes such devices a leading disruptive technology in this new field.

I keep a greeting card that someone gave me long ago on the wall of my office because it makes me laugh. It's a cartoon of a blackboard in a laboratory, on which someone has written, "If I were a scientist in a lab, I would shout EUREKA! every once in a while, just to boost morale." The only thing I would change about this today is that scientists don't say EUREKA! anymore, they say HOLY SHIT! That is what we all said the day the drug in the rat's brain turned off the body's inflammation because of the vagus nerve.

After training to become a neurosurgeon for almost ten years, then working as a neurosurgery attending for fifteen more, I retired from clinical practice. This enabled me to place all my efforts on my laboratory and pursue the mission of the Feinstein Institute that I lead:

Producing knowledge to cure disease.

I am convinced that bioelectronic medicine has the potential to revolutionize healthcare and improve the lives of countless people around the world. Although it's not yet in textbooks or widespread practice, bioelectronic medicine is here to stay. It's in my lab and others like it around the world, in clinical studies, and in therapeutic devices already approved by the FDA. It has already changed many people's lives for the better. The rest of this book tells the story—or many stories, spanning many centuries—of what makes this new field possible and how helpful it is and could be. In my lab, morale is high, and the "eurekas" are real.

FROM BENCH TO BEDSIDE

Scientific research and biomedical engineering are leading to new vagus nerve stimulating devices that can be used to treat inflammation. Clinical trials in the United States and Europe are showing impressive results. Some patients, like Kelly Owens, enter a trial after spending decades suffering from joint pain, fatigue, muscle stiffness, abdominal distress, and hospitalizations from complications of rheumatoid arthritis and inflammatory bowel disease. With regular vagus nerve stimulation, many of these patients are enjoying normal, healthy lives, free of their disease, while taking no medications.

The last and final step will be the adoption of vagus nerve stimulators and other bioelectronic therapies into medical practice for everyone who can benefit. In the history of medicine, the uptake of a truly new idea can take years or even decades. There are many good reasons for this, including the conservative decision-making nature of physicians that starts with first, do no harm. You don't want your doctor chasing every shiny new object. But other, less benign forces may also hinder the adoption of vagus nerve stimulation because this new idea has the potential to disrupt the pharmaceutical industry. Today, anti-inflammatory medications represent a substantial proportion of total revenue and global drug sales.

Greater awareness of the power and practicality of vagus nerve stimulation, combined with increased demand, will accelerate the adoption of this revolutionary approach to medicine. While medical progress often moves slowly, this book aims to equip you with the knowledge to discuss your vagus nerve with your doctor today. This will enable you to access new, scientifically proven therapies as soon as they become available. As we peer into a clear and present future of precise and personalized vagus nerve care, you may well ask:

- Is there a vagus nerve therapy for my diagnosis or what's bothering me?
- Can I stimulate my vagus nerve without a surgical implant?
- What do I need to know to choose a vagus nerve therapy?
- What are the questions I should ask my healthcare providers?
- Could I join a clinical trial if the therapy hasn't been approved yet, as Kelly Owens did?

- What if I just want to feel happier, stronger, and live longer, content in my intentional day-to-day relationship with my vagus nerve?

These are all good questions. Let me walk you through the options, so you can make informed decisions on the path of health and harmony with your vagus nerve. Its song is your song, after all.

2

The Great Nerve Reveals Itself

Lucia Galvani told her husband Luigi that the frogs' legs prepared for the meal seemed alive on the copper-wire. . . . The ensuing century's analysis resolved this aspect of the spirits of the anima into transient electrical potentials travelling the fibres of the nervous system. They were no longer "spirit," but were become a physical event describable under energy.

—CHARLES SHERRINGTON

You might not have known you have something called a *vagus nerve* before you picked up this book. Or you might be inundated with vagus-related tips from many directions, depending on your relationship with the social media algorithms. So, let's begin with a simple, unifying question:

Do you want to be healthier and happier?

If you said no, because you have achieved a state of health and mood that cannot be improved, then please accept my congratulations. You are in a fortunate group of individuals who (whether or not you know it) have learned how to tune their vagus nerve and harness its powers for optimal health. You are likely to enjoy a longer lifespan, suffer less illness, and experience more emotional resilience compared to those whose vagus nerves are not as well tuned.

If you said yes, that you would like to be healthier and happier, you are not alone, obviously. However, less obviously, there are abundant opportunities for you to work with your vagus nerve and capture the benefits to your physical and mental health. The structure and function of this nerve were shaped and honed by millions of years of evolution, its purpose to optimize your whole body's responses to life's challenges, large and small, keeping your systems in balance and your organs working in harmony and bolstering your cognitive abilities and sense of well-being. Increasing your vagus nerve tone may enhance your resilience, improve your mood, and boost your overall well-being.

Vagal tone is the term for the activity of the vagus nerve. It is a reflection of the neurological activity of the parasympathetic nervous system, and there are things you can do to improve it. Things like physical exercise, meditation, breathwork, and cold exposure. Improving vagal tone is linked to a diverse array of health benefits, including a reduced risk of heart disease, stroke, and diabetes, as well as improved mood, cognitive function, and sleep quality. In other words, health and happiness.

How can two hundred thousand nerve fibers traveling up and down in one bundle on each side of your neck between your brain and all your organs control so many vital outcomes, making the difference between living healthier and happier, or not? Why do subtle modulations of the signaling activities of these fibers make such a difference? What secrets have scientists unlocked about the vagus nerve's role in health and well-being over two thousand years of research? We have answers to these questions because scientists have been studying the vagus nerve for two thousand years.

We are besieged with instructions about how to eat, sleep, breathe, work out, meditate, relax, bathe, and cleanse while simultaneously maintaining work-life balance to live long and prosper. It's sometimes overwhelming. We're constantly bombarded with conflicting advice streaming by on social media, television infomercials, and podcasts, and it's hard to know what to prioritize. There doesn't seem to be enough time in the day to be healthy. After all, most of us have obligations and responsibilities that come before working out, cooking organic food, and taking ice baths, yet we all know that little matters more than health and happiness. But there are solutions, and a day when doing the healthy things seems overwhelming is a good time to return to basics. It's always best to understand the nature of a problem before we try to fix it.

This chapter will introduce you to your vagus nerve through some of the key experiments and discoveries people have made about it. In the process, you'll get a sneak preview of the kinds of everyday practices that help the vagus nerve continue to do its amazing work. (More on these in part 3.) As you get to know the nature of your vagus nerve, the far-reaching benefits of keeping it in tune will make more and more sense.

For those seeking more detailed insights into the physiological mechanisms of the vagus nerve and the research that revealed them, hundreds of scientific treatises are available. I cite many of these in the Notes. Because some are perhaps difficult to understand unless you are a specialist in physiology, anatomy, neuroscience, immunology, psychology, or some other medical field, I have tried in this book to explain the main points in (hopefully) understandable language. I hope to bring the vagus nerve to life as if you are seeing it for the first time through the eyes of the scientists, doctors, and inventors who first revealed some of its secrets.

If you're eager to cut to the "how-to" part of the chase, you can skip ahead to part 3, but my caution is that you will get much more out of the practical applications if you first understand how the vagus nerve works and why it is truly great.

EXPOSING THE VAGUS NERVE

Claudius Galenus was the original physician-scientist. Better known as Galen, he lived to be an old man twenty centuries ago, with a highly successful medical practice at a time when practicing medicine and dying old were both difficult to do. His life spanned the peak of the Roman Empire and the beginning of its decline, the last of the Five Good Emperors and some of the not-as-good-ones that followed, several of whom Galen personally attended. Born into wealth, the son of a prosperous Greek architect who died when Galen was nineteen and left him everything, he lived a relentlessly curious life. On some days, he did surgery on wounded emperors and gladiators, and on other days, he plied his skills on their prized pets and farm animals. Believing each incision offered him a unique new "window into the body," Galen peered deeply, trying to understand the secrets of the brain, the mystery of its nerves, and their role in the mechanisms of life.

There was much to discover. Galen's formal education from the treatises of Hippocrates taught him that the brain was the source of human consciousness, thinking, and willpower, but nobody had figured out what individual nerves do or that they were all part of the same system. Centuries earlier, Aristotle had proposed that nerves were controlled by the heart, where they were thought to originate. But Galen carefully traced the vagus nerve and discovered its origination in the brain stem.

Because dissecting human cadavers was illegal in Galen's time and place, his most famous experiment involved the vagus nerve of a pig. Believing animal organs and functions mirrored their roles in humans, he sought to unlock their mysteries through hundreds of dissections in pigs, sheep, baboons, and monkeys. During these studies, he encountered the vagus nerve—and followed its lead through a series of discoveries that reshaped his world's understanding of the nervous system.

Imagine the crowd huddled close around the table as the surgeon lays bare the muscles, arteries, and nerves of the living pig's neck, an event part educational experience and part live public service announcement. Watch as he follows one of the vagus nerve's branches in the neck as it descends away from the brain, parallel to the carotid artery. Notice how it makes a sharp U-turn upward and dives into the structure we call the *larynx*, positioned in the center of the throat behind the thyroid cartilage recognizable from the outside as the Adam's apple. Nobody in the audience knew anything about how nerves or the larynx worked, but this was about to change.

As the art and science of anesthesia were a dream centuries in the future, all the surgeries Galen performed, whether on humans or animals, were painful and loud. But after locating the nerve that branched away from the main vagus and tracked back upward in the squealing pig, Galen cut it. Then silence, though the pig was still very much alive.

With one flick of his knife, Galen showed the world two things.

First, he proved the brain controls behavior by sending (then unknown) signals through nerves. It would be a long time before we'd start to understand that the signals go both ways, but Galen's brain-to-body view of the nervous system, though technically ancient, was revolutionary and surprisingly modern. Before Galen, we had Aristotle's model, which held the brain as merely a cooling organ for the blood and the heart as a conductor of the body's functions through its arteries. Galen centered the brain in our understanding of the body like Copernicus, thirteen centuries later, centered our solar system on the sun.

Second, because that branch of the vagus nerve that silenced the pig when he cut it terminated in the larynx, he proved that the larynx was the source of vocalization. Today we know that within the larynx, the vocal folds, or vocal cords, modulate airflow, allowing people to make words and music, and pigs to oink, grunt, and squeal. Through intricate movements and tension adjustments, the vocal cords create a dynamic interplay of vibrations that give rise to the diverse range of sounds and melodies we express (or try to, at least). The vagus nerve modulates all of this by sending signals up into the brain about the status of the larynx's muscles and soft tissues and by transmitting signals back down from the brain that coordinate the words and songs. Some vagus nerve fibers send signals up to the brain, and others send signals down to the body. A massive quantity of input and output data, between the body and brain, or in this case, voice box to brain and brain to voice box, travels via the two hundred thousand fibers of the vagus nerve.

We've come a long way since Galen, and while I am not a historian, from my twenty-first-century physician-scientist's vantage point, I have seen new vagus nerve discoveries leading to new therapies. I know from experience what it's like to operate inside the body and brain. I've seen human beings in various states of physiological compensation and decompensation. I've had intimate encounters with the vagus nerve's anatomy and its functions. I've tended and befriended human beings suffering from illnesses and living with the results of successful or failed surgeries and other treatments.

As an experimentalist in a laboratory, I've had the privileged opportunity to dissect and reveal in exquisite anatomic, cellular, and even molecular detail some of the secrets of life. We listen to the vagus nerve and record it, gather data, and tell the world what we've learned. Because the questions we're asking are about the interactions between complex living systems, which cannot be answered in test tubes and petri dishes, we study mice.

To do this, we spend millions of dollars a year on murine welfare, housing them in a warm and clean environment offering an abundant, healthy diet. Our work at the Feinstein Institutes meets the highest standards of ethical oversight guided by the U.S. Department of Agriculture and New York State. Every experiment we propose is reviewed by an independent committee responsible for the ethical treatment of animals, including the kind of anesthesia we use and how we draw their blood. All this is, as it should be, the responsibility my colleagues and I accept in the hopes of discovering new cures and treatments for people needing them most.

Nearly two thousand years after Galen, imagine you're a witness to *this* scene: We're in my laboratory, in a five-story modern glass and brick building on a suburban corner of Manhasset, New York. We swipe into an air-locked, brightly lit, pressure-and temperature-controlled hall of mouse lodging and operating rooms. After gowning up (you, too) in paper robes, shower caps, shoe covers, masks, and gloves, we enter the housing room lined with stainless steel shelves displaying clear-sided portable compartments with soft, mulchy bedding. The room and everything in it is spotlessly clean, and it smells pleasant, warm, and yeasty. A twelve-hour light cycle complements the ambiance, controlled to accommodate the rodent preference for being awake and feeding when most researchers are asleep: A timer lowers the lights so that the murine "night" corresponds with our day. The surgeon studies the labels on a row of compartments until she finds the one she's looking for and slides it out from its shelf. Our shoe covers make a swishing sound down the hall as we walk to the operating room, where a surgical table is prepped for microsurgery.

The surgeon gently lifts one of the mice from its container into a closed Plexiglas box fed with halothane anesthetic gas by a tube. When the mouse is sedated, in ten or fifteen seconds, the surgeon scoops it out, laying it onto a padded table on its back and guiding its cone-shaped snout into a small cone-shaped nozzle that will deliver precise amounts of anesthesia throughout the procedure. As the mouse breathes steadily into the cone, the surgeon daubs a small amount of Nair on its throat and upper chest, waits ten or fifteen seconds, and carefully wipes away the fur where she will make a tiny incision, so tiny that she will view what she's doing through a microscope. She rubs alcohol on the exposed skin to sterilize it and swings the microscope into place.

A flat-screen monitor connected to the microscope shows the scalpel cutting the skin slightly left of the neck's center, a few millimeters long and one millimeter deep. The cut is just deep enough to reveal the strap muscles, those neck muscles you can feel to the side of your own carotid pulse, adjacent to your Adam's apple. The surgeon nudges apart the mouse's strap muscles to expose the mouse's carotid artery, encased with the vagus nerve in a fibrous tunnel that the surgeon carefully incises to lay bare the nerve. Although it is only the thickness of a human hair, on the monitor the white vagus stands out next to the carotid artery's beating red. Using microtweezers, the surgeon gently lifts the delicate nerve and slides an electrode underneath it, pausing to rest the nerve for a few minutes before she turns on a computer that controls the electrode.

The electrode, a product of Silicon Valley technology, is capable of both stimulating the mouse's vagus nerve with calibrated electrical pulses and recording the billions of electrical signals flowing back and forth between the body and the brain. As you watch, the surgeon's computer captures the vibrations normally conveyed by vagus nerve fibers onto its hard drive. It also records the changes in these signals brought about by electronic stimulation. As if we're in a vagus nerve recording studio, the signals we capture can be amplified, filtered, and converted into a digital format, just like music.

When the recording session is done and the electrode removed, the surgeon closes the tiny incision with a single stitch and draws a blood sample to analyze later in the lab to reveal the effects of vagus nerve stimulation. Then she lifts the mouse to a recovery compartment, where it will awaken before long.

In these everyday lab experiments, I think of Galen's pig. Centuries and generations of physician-scientists later, what was then a brand-new idea still rings profound: *The brain listens to the body through the vagus nerve and then talks back.* Twenty centuries ago, Galen's vagus nerve discovery pointed to a new way of thinking about how the brain uses nerves to control your body. And even today, it seems the vagus nerve still speaks to us, to my colleagues and me, challenging us to reveal its other hidden mysteries that influence your health and happiness.

YOUR VAGUS NERVE AND YOUR BEATING HEART

Since it is the nature of your autonomic nervous system to function without your conscious knowledge and we cannot directly measure the vagus nerve's activity without surgery, how do you know if your vagus nerve is being stimulated? A simple answer is that you know it is being stimulated when your pulse slows.

By the middle of the nineteenth century, the scientific world, fascinated by electricity, was obsessed with a host of newfangled electric batteries, generators, Leyden jars, and other devices to stimulate the muscles and nerves of laboratory animals and humans, including the scientists themselves, prisoners, and even freshly guillotined corpses. The consensus opinion from widespread results indicated that applying electricity to nerves caused the attached muscles to contract, offering an explanation for how signals were carried in the nerve Galen cut centuries before. As word of these results spread, a new idea took root: that nerves transmitted electrical signals to intensify the actions of muscles and organs. Then a pair of physician-scientists working in their lab in Germany electrically stimulated the vagus nerve, and things got really interesting.

By the time the Weber brothers, Ernst Heinrich and Eduard Friedrich Wilhelm, began their experiments, it was known that the vagus nerve sent branches to the heart. But when they turned on an electric current (by rotating a copper disk immersed in a magnetic field to generate electricity—a Faraday device) and applied it to the vagus nerves of frogs and other animal models, they were astonished to observe that the heart rate did not intensify. It actually *slowed down*. Afraid they had somehow made a mistake, they repeated their experiment, stimulating the vagus nerve many times. Each time, it produced the same result. They found that by turning the electric current up high enough, they could even bring the heartbeat to a complete standstill, after which it resumed beating normally again. This was good news for the frog, and for the millions of people who have benefited from nerve stimulation ever since.

In science and medicine, every new discovery begets a new enigma, and the Webers' discovery delivered a big one. Nerves don't only excite; they can also inhibit function. But how? How does an electrical stimulus in the vagus nerve produce its slowing effect on the heart? And what purpose does this serve? The answers to these questions appeared seventy-five years later when the Austrian pharmacologist and researcher Otto Loewi found a clue in his sleep.

After a long day spent thinking about how nerves function, Loewi's sleeping mind painted a dreamscape of the vagus nerve delivering cryptic signals to the heart, even going so far as to conjure the outlines of an experiment. He bolted (so to speak) to his lab while the dream was still clear. There, in the predawn quiet, he carefully removed the beating hearts from two frogs, keeping the vagus nerve attached to one heart while removing it from the other. He placed each heart in a saline bath, one still attached to the vagus nerve, the other disconnected.

When Loewi electrically stimulated the vagus nerve on the first heart, as the Weber brothers had done years before, the heart's beat slowed, this time as expected. Then, as Loewi had dreamed, he transferred liquid from this first bath into the second and watched as the second heart, with no nerve connected, began to slow as well. He realized that *electricity* in the vagus nerve caused it to release *chemicals* that slowed the heart—a then shocking observation that explained, at the most basic level, how the nervous system actually works.

Loewi's discovery that nerves convert electrical signals into molecular ones meant that molecules released by the nerves, not electric current directly, are responsible for controlling the function of the organs receiving the neuronal inputs. He called the elusive chemical released by electrical stimulation of the vagus nerve that slowed the heartbeat *vagusstoff*, which means what it sounds like: vagus material, or vagus stuff. Today we call it *acetylcholine*, the first identified neurotransmitter, and we understand that among its many effects on the body's cells, it also suppresses cardiac pacemaker cells, which slows the heart.

Loewi repeated his experiments hundreds of times, stimulating not only the vagus nerve to slow the heartbeat (stimulating "rest-and-digest") but also stimulating the sympathetic nerves to the frog heart that accelerate the heartbeat (stimulating "fight-or-flight"). Transferring liquid from the bath of an innervated frog heart beating faster to a denervated heart in a second bath, he observed the second heart racing. He discovered that applying electricity to sympathetic nerves stimulates the release not of acetylcholine, but of a different molecule, which speeds up the heart. Today we call the molecule released by sympathetic nerves to accelerate the heart the neurotransmitter *norepinephrine*.

Acetylcholine and norepinephrine are the first two words that researchers discovered in a complex language of electrical stimulus and chemical neurotransmitter responses that direct our cells, muscles, and organs. More than one hundred additional neurotransmitters would be identified in time, each with its own frequency in the harmony of the nervous system.

Recalling my medical school professor Dr. McNally's summary

of the autonomic nervous system, it turns out that one major difference between the fight-or-flight and rest-and-digest responses lies in the neurotransmitters they release. Sympathetic nerves produce norepinephrine, which tends to speed things up, and the parasympathetic vagus nerve produces acetylcholine, which tends to slow things down.

Scientists and physicians studying the regulation of normal heart rate since the Weber brothers and Otto Loewi learned that a resting human's heartbeat can vary across a range between sixty and one hundred beats per minute, depending upon the age, health, and fitness of the person. Your pulse is influenced by a balancing act between signals in your vagus nerve, which slow your heart, and signals in the sympathetic nerves, which speed it up.

Generally, research has found that a slower pulse within the normal range is a good thing if you want to live longer and healthier. One study from Framingham, Massachusetts, assessed the impact of resting heart rate on mortality over a thirty-year follow-up period in 5,070 participants who had no cardiovascular disease at the beginning of the study. The analysis, which was based on a total of 1,876 deaths, revealed that for both men and women across all age groups, there was a progressive increase in mortality rates from all causes, as well as from cardiovascular and coronary diseases, that corresponded with higher resting heart rates (as recorded every two years). The people with slower pulses were significantly healthier and lived longer.

A second, even larger study from France evaluated the impact of high heart rate on mortality among 19,386 French adults, aged forty to sixty-nine, from 1974 to 1977, with follow-up until 1994. The results were stunning because they showed that a faster pulse in a population is an independent risk factor for noncardiovascular mortality in both sexes and for cardiovascular mortality in men. In other words, once again, people with slower pulses and, therefore, more vagus nerve activity were significantly healthier and lived longer. Checking resting pulses is the simplest, cheapest measure to determine the statistical longevity of a population and doesn't require blood tests or other more invasive, expensive diagnostics.

If you put down this book, check your pulse, and find that it's on the high side, let me reassure you this does not mean you're doomed. In fact, some people with rapid heart rates live to be more than one hundred years old. How does this square with the studies? I like to explain it with the analogy that if you're holding a winning lottery ticket, your chances of winning are not one out of 270 million; they are 100 percent. Statistics derived from a population study do not predict the specific risk to a specific individual.

Nonetheless, we do know that a slower resting heartbeat is the result of a healthy vagus nerve and is, statistically, a good thing. Vagus nerve activity, or vagal tone, is an integrative function that reflects your prior inputs from exercise, among other behavioral, environmental, and genetic factors over many months or years. Therefore, a slower pulse for your age and health could reasonably indicate progress toward your goal to be healthier and happier, as it suggests you are living in ways that chronically stimulate your vagus nerve. We'll talk more about strategies to do this in part 3.

HOW YOUR SYMPATHETIC AND PARASYMPATHETIC SYSTEMS COOPERATE

It was long thought that fight-or-flight and rest-and-digest responses achieved balance in the body by being mutually exclusive. If one was "on," scientists reasoned, the other must be "off." When your heart rate needs revving up for exercise, say, it was assumed that your vagus nerve must be inhibited during those periods. We now know this assumption is wrong. Recent research has revealed a more dynamic relationship between these two systems, where they can both be active to varying degrees depending on the situation.

To grasp the bodily effects of the autonomic yin and yang in action, imagine that you're jogging along at a comfortable pace on a quiet street when a large dog comes racing at you from a stranger's yard, snarling, teeth bared, ears forward, with a stripe of fur bristling along its spine. Immediately, your sympathetic nervous system is activated, preparing you to either fight or flee. Your heart rate increases, your blood pressure rises, and your pupils dilate. If you choose to run, your sympathetic nervous system provides the energy and strength needed to escape (hopefully). If you decide to fight, the adrenaline and glucocorticoid steroids coursing through your veins focus your mind. They make time stand still and put you in an optimal state for physical prowess and self-defense.

But then the dog's owner appears, calls out, and the well-trained animal stops, turns around, and retreats to its person. The danger has passed. As your sympathetic activation subsides, your parasympathetic nervous system takes over. As your heart rate decreases and your blood pressure drops, your body returns to a state of calm. These fight-or-flight and rest-and-digest responses are both essential because life's challenges can be extreme and because both are required to maintain balance in the body. Recent research reveals they are in a two-system model of cooperation rather than simply being at odds. They work together to establish a delicate balance, constantly adjusting to maintain homeostasis and ensure our survival.

High vagal tone indicates a strong parasympathetic influence, which is usually associated with a prevailingly relaxed, calm state. Conversely, low vagal tone with high sympathetic activity is typically associated with stress or anxiety. We sometimes gauge vagal tone by measuring heart rate variability (HRV), which is the variation in time between each individual heartbeat, as measured by analyzing the spikes on an electrocardiogram collected over a period lasting from an hour to a full day. Each time the vagus nerve sends a "slow down" signal to slow the heart, it prolongs the time until the next heartbeat. So high HRV, indicative of high vagal tone, means that the vagus nerve is more activated, producing a body that can efficiently shift from states of stress to relaxation and, often (but not always), a slower resting pulse. This is an indicator of good health. Conversely, low HRV or low vagal tone can be a sign of chronic stress or illness and is linked to higher risks of cardiovascular diseases and other health problems, and usually a faster resting pulse.

Let's go back to our jog. If, as many textbooks still say, exercise increases sympathetic activation, which in turn switches off your parasympathetic system, including your vagus nerve, then jogging and other types of aerobic exercise wouldn't seem to benefit your vagal tone. However, important new results indicate that exercise gives your parasympathetic system a workout too. This research challenges the traditional textbook view and opens up new possibilities for understanding the complex interplay between exercise, the nervous system, and overall health.

One recent study that sounds like it could be the basis for a *Far Side* cartoon observed sheep on a treadmill (the study was done in

New Zealand). First, the researchers determined that the optimal sheep exercise regimen was 18 minutes, up to a maximum speed of 2.5 kilometers per hour on a 15 percent incline. This increased the sheep's cardiac output four- to fivefold—a reasonable effort for animals that evolved to roam for miles across hill and dale in order to simply survive. The scientists implanted specialized catheters to probe, monitor, stimulate, and record the activity of each sheep's heart, vagus nerve, and blood flow during exercise. They discovered that, contrary to expectations, the vagus nerve was not "off" as the treadmill whirred and the sheep trotted; vagus nerve activity actually increased during the workout.

To understand why this occurred, they measured blood flow through the coronary arteries supplying the heart and discovered that the increased vagus nerve activity produced increases in the coronary blood flow. This increased blood flow to the heart muscle enabled it to meet the increased demands of exercise. The enhanced vagus nerve signals delayed the onset of the next heartbeat, giving the heart more time to fill up with oxygenated blood, resulting in increased oxygen delivery to the heart and the body with each beat. This study showed a mutually synergistic benefit, rather than opposition, between the two sides of the autonomic nervous system.

If you are healthy, your vagus nerve is never "off" but always working to keep your body in dynamic balance and you as healthy and happy as you can be, whether you're at rest, exercising, or facing stressful situations. Regular exercise improves your overall vagal tone, a sign of enhanced mind-body communication essential for overall health and well-being. We will return in part 3 to talk more about exercise, vagal tone, and cooperation in your nervous system.

THE BRAVE NEW WORLD OF OPTOGENETICS

Advances in molecular biology, neurophysiology, and electrophysiology have given us modern windows into the body that Galen, the brothers Weber, Otto Loewi, and other early researchers couldn't imagine, revealing the workings of the vagus nerve in further and deeper detail and letting us learn how every one of its two hundred thousand nerve cells carries information up to your brain and back to your body.

Each of your two hundred thousand vagus neurons is a cell with a long projection, the axon, that produces tiny electrical spikes called *action potentials*. These spikes flow along the nerve fiber, conducting a current of information in an intricate twining of biology and physics. At rest, the neuron is a silent vessel brimming with potential energy. But when it receives the signal to activate, the neuron erupts into life. Channels in its cell membrane open, allowing a flood of positively charged sodium ions to rush in. This influx changes the electrical charge within the neuron, generating a voltage spike—the action potential. Then, a wave of electrical excitation speeds along the nerve, triggering a chain reaction of sodium channels opening sequentially along the nerve's length, propagating a stream of action potentials. The once quiet vagus neuron has become a buzzing, humming messenger, its electrical signals hurtling toward the end of the line (the *presynaptic terminal*). The arrival of action potentials at the end of the axon causes neurotransmitters to be released from the terminal into the small space (the *synaptic cleft*), separating the nerve ending from its target organ or another nerve cell. A brief but crucial silence, a refractory period, follows, allowing the neuron to reset itself for the next wave of information.

With new technologies, we are able to begin teasing apart this tangled web of connections and link them to behaviors. One powerful tool, called *optogenetics*, combines the precision of laser optics and the intricacies of molecular genetics. In a leap akin to Otto Loewi revealing how nerves convert electrical signals into chemical ones, we are now genetically engineering neurons in mice to express light-sensitive proteins, or opsins, originally derived from sunlight-loving algae, in order to use laser beams to activate neurons. In their natural environment, opsins enable algae to orient themselves toward sunlight. Once introduced into neurons, opsins make those brain cells responsive to light. By deploying laser light beams of a specific wavelength that interact with the opsins, we can either stimulate or inhibit specific neurons in the brain with unprecedented precision and watch what happens. This cutting-edge technology is revolutionizing neuroscience research and offers the potential to develop novel treatments for neurological disorders.

Optogenetics gives us the power to selectively activate a target set of neurons out of the billions available, allowing us to observe the resulting behaviors. It is like the difference between playing a delicate melody on a piano and sitting on the keys—if the piano had a hundred billion keys. We can literally see the connections between the activity of specific neurons in a discrete brain region and the specific physiological, pathological, and behavioral functions they control in the body. Several labs, including my own, are taking up these new tools to study the vagus nerve.

Researchers at Harvard Medical School have developed another molecular toolkit that, when combined with the potent lens of optogenetics, can be used to explore the vagus nerve in mice, bustling with its four thousand to six thousand neurons (simpler animal, fewer fibers). Out of this complex neuronal constellation, they have focused their efforts on mapping a precise cohort of a mere few hundred stars and made a surprising discovery: Two separate groups of neurons in the vagus nerve, numbering perhaps a few hundred in each group, have contrasting impacts on the breath. When the researchers shone a laser light on the brain to optogenetically stimulate neurons in one group, the mice stopped breathing while stuck in a deep exhale phase. But stimulating the other group of neurons caused the mice to take rapid, shallow breaths. These breath-controlling vagus nerve fibers did not affect heart rate or stomach pressure, other functions regulated by the vagus nerve. This means that specific groups of vagal neurons serve their own specific functions; stimulating one small part or segment of the vagus nerve can produce completely different responses from stimulating another small part.

So when someone says to me, "I want to stimulate my vagus nerve," I say, which part? Because that matters.

Deep breathing is one way you can voluntarily stimulate your own vagus nerve. Try it now. Take a deep diaphragmatic breath by imagining you are inflating the lower portion of your lungs near your belly button like a balloon, followed by a gradual exhalation as you pull in your belly button. Aim to inhale through your nose this way for a count of three, then exhale slowly through your mouth for a count of seven. Then repeat this cycle twelve times, equivalent to twelve complete breaths (or two minutes), and you are on the way to enhancing your quality of life.

It's that easy. You just activated your parasympathetic nervous system, slowed your pulse, enhanced your heart rate variability (HRV), and increased your vagal tone by stimulating some of your vagus neurons—probably a few hundred fibers on the inspiration (inhale) and a few hundred more on the expiration (exhale), although we're not really sure. We do know, and we'll explain more as we go, that you probably feel better after deep breathing, and there are good reasons for it. As you'll see in part 3, by making habits that combine diaphragmatic breathing with regular aerobic exercise (brisk walking, biking, swimming, or running), you can further cultivate the long-term benefits of a healthy vagus nerve.

HEALTH, HAPPINESS, AND YOUR VAGUS NERVE

In Galen's time, the nerve we call vagus was known as the *pneumogastric nerve*, named then for the path it traces to the lungs (*pneumones*) and stomach (*gaster*). After delving more deeply into its functions, Galen thought it needed a name change. Out of respect for its voice-giving and life-sustaining importance, he christened it "the great nerve," which was how it was known for more than 1,400 years.

Fifteen centuries after Galen, in the early 1600s, physician-scientists perfected dissection procedures in animals and human cadavers and continued to trace the anatomy and study the functions of the great nerve. They cut it in various places, forming new observations, both interesting and grim. They bisected the great nerve near the brain on both sides (left and right), typically at the neck, and found this was enough to end the lives of dogs, cats, pigs, frogs, and other laboratory subjects. They traced its branching course from the brain through the body, a spider's web connected to all the organs in its path. They confirmed the nerve's essential role in sustaining life. Impressed by the nerve's long reach and staking a new claim, one of these scientists in Denmark,

Caspar Bartholin, chose to rename the nerve again, calling it the *wandering nerve*, or *nervus vagus* in Latin.

Wandering doesn't seem quite right to me. The more we learn about the vagus nerve, the more we understand that it knows exactly where it's going and what it's doing. Modern science has revealed it to be precisely wired and finely tuned, an ally in health, happiness, and as we will see in its relationship to the immune system, in healing. The nature of the vagus nerve is far from aimless—Galen and the early scientists who followed him were already on to this. The purposefulness of the nerve captivated their scientific curiosity, as it does mine.

Maybe "vagus" poetically reflected the winding lines they had so far traced, or maybe it inadvertently expressed the scientists' own uncertain understanding of its great powers. There was much left to learn. There still is. In any case, the name vagus held—but the more we learn, the greater I know it to be.

Two thousand years of vagus nerve science continues to evolve. In weekly meetings in my lab, we review our latest experimental results. Thinking about what it all means, my lab codirector, Sangeeta Chavan, and I, together with our teams, look for answers buried in the vagus, searching for the clues our data tell to solve its mysteries. Like others in their labs around the world, we keep probing for more clues. We listen to the vagus nerve as it transmits its messages back and forth between the body and brain, one fiber at a time. Because of the scientists, physicians, and innovators who paved the way before us, and what we have learned since, we are confident now that these studies will lead to healthier and happier lives for others. And perhaps you.

3

Your Body's Healing Reflexes

The constancy of the internal environment is the condition for free and independent life.

— CLAUDE BERNARD

Her name was Janice, and she was only eleven months old when she crawled across the kitchen floor one evening while her grandmother was immersed in the rhythm of preparing dinner. As her grandmother turned to drain a pot of boiling water, she tripped over Janice, and a cascade of scalding liquid landed on the helpless child. Less than an hour later, thirty-nine years ago, I found myself in Janice's world, in the burn unit on the seventh floor of the Payson Pavilion of the New York Hospital overlooking the East River, as I inserted a maze of catheters into her femoral arteries and veins to provide her with the fluids necessary to sustain her tenuous existence.

Burn units are brightly lit, warm, humid, and full of suffering. A stench forms from the mingling of assorted malodors—human sweat from hardworking staff entwined with the pungent antibiotic Silvadene, a healing balm we apply to the burn wounds, which emanate their own putrid fumes. In this fetid atmosphere, we must make cold calculations, using formulas estimating mortality from the patient's age and the volume of burned and injured tissue. In Janice's case, the shadow of death loomed large, its probability exceeding 90 percent.

But for three and a half weeks she defied these calculations, her spirit persevering even through pneumonia and a failing liver and kidneys. As she continued to rally, we hoped maybe, just maybe, she would beat the odds. We began to whisper of her future, of days beyond the pain, of possibilities unthinkable since the accident becoming thinkable again. Maybe someday she could go home.

It was in this soft glow of hope that I stood in the doorway to her room, watching a nurse gently rock Janice in her arms while feeding her from a bottle, back and forth and back and forth until suddenly something was wrong. I saw Janice's eyes roll back in her face (which had escaped burning). On the screen over their heads, the numbers flashed as her blood pressure dived and her heart rate launched into a chaotic race. As her lips turned blue, and bluer still, Janice stopped breathing.

I rushed into the room, cradled Janice in the hollow of my left arm, sealed my mouth over her nose and chin, and breathed gently into her lungs as I rhythmically pressed one finger to her fragile sternum. The nurse summoned the entire staff to Janice's code. Within minutes, her room was thronged by some of the finest pediatricians, intensivists, and surgeons in the world.

Baby Janice was intubated, placed on a respirator, cardioverted with electric shocks, and ultimately, she received an intracardiac pacemaker floated into her right atrium through her jugular vein, something that until that day had only been done one or two times in small children at our hospital. Everything, even that rare procedure, went flawlessly during the hour-long code, yet despite our efforts, her heart stayed silent, and we could not restart it.

It was my responsibility to tell Janice's mother. I can still hear her scream before she fell quiet, a stunned silence filling the room as if the whole world stopped breathing with her daughter. Then she dropped like a sack onto the floor. We gently lifted her up onto a couch, lowering her head and elevating her legs. After she came to and was surrounded by family, I exited to the nursing station to write the "death note" in Janice's chart.

In the months that followed this loss, my thoughts often wandered back to the haunting question of what had led Janice to that precipice. We all understood then that burn patients walk a fine line with infection. We were fully aware of the toxins created by microbes, the best known and studied being lipopolysaccharide (LPS), a lethal toxin expressed by *E. coli* and other bacteria. And we knew that LPS triggers a physiological failure with plummeting blood pressure, erratic heartbeat, the blackening of necrotic organs, and the cessation of breath. It was as if Janice had been poisoned with LPS, yet there was no sign that she had any significant infection, no trace of toxins.

If there was no infection and no bacterial toxins, then what had triggered the collapse? I knew a piece of the puzzle was missing. And I knew that whatever was missing must lie hidden in plain sight, in the intricate interplay of life, death, and inflammation.

YOU ARE YOUR REFLEXES

As you read this, your brain is actively engaged in the tasks of scanning the text and converting the printed words into a voice heard only by you, inside your head. You can hear this internal speech because your brain handles written words in your auditory cortex, which is responsible for processing and interpreting sounds. But as you trace the vagus nerve from Galen's pig surgery to my lab, or try to understand why Janice died, your brain is also multitasking, managing myriad other functions that keep you alive and healthy and capable of reading. Nearly all the brain's tasks unfold beneath your conscious awareness. For example, unless you intentionally shift your focus now from this page, you are not aware that the sensory nerves in the soles of your feet are relaying information to your brain about the pressure exerted by your socks against your skin, although should you decide to pay attention to them, you can become aware of your foot sensations. No matter how hard you try, however, you cannot consciously attend to all the sensations arising in most of your body.

Sensory nerves surround the cells in your body, including all your organs, and even if you want to, you cannot feel the signals that control your internal organs and keep you healthy. You cannot feel how the glucose level in your liver prompts sensory nerves traveling up to your brain through your vagus nerve to coordinate the need for the pancreas to produce your insulin. Nor can you feel the sensory nerve signals in your vagus nerve informing your brain about the pressure of air in your lungs or levels of carbon dioxide in your breath, or the signals coming back from your brain through the motor nerve fibers in your vagus that modulate the activities of those same organs. Your body temperature, heart rate, and blood pressure, along with oxygen, hormone, and glucose levels and millions of other vital data points, are constantly relayed from your body to your brain through your vagus nerve.

You are oblivious to these processes. Meanwhile, your nervous system uses the power of its one hundred billion neurons and its six hundred trillion synapses to process these vast amounts of sensory data—its input. Input and output, sensations and responses to those sensations are happening every second you're alive.

I've been thinking about this for decades, and the question that keeps me awake at night when my vagus tone would benefit from sleep is, how? *How* do all these systems work without our having to think about them?

The answer, distilled to its essence, is reflexes.

You are likely familiar with your own knee-jerk reflex. I'm not talking about a reaction to someone cutting you off in traffic or to having your political buttons pushed. I mean the literal knee jerk activated by your doctor as you sit with your legs dangling over the end of an exam table. Also known as the *patellar reflex*, it starts when a rubber hammer taps against your patellar tendon just below your kneecap (*patella* is another name for kneecap). This little thump stretches the tendon and stimulates sensory nerves that specialize in responding to tension and stretch, producing a series of action potentials in the sensory nerves that travel to the spinal cord, where neurons there send output via motor nerves back to the quadriceps muscle, which contracts. Before you are even aware anything happened, your foot surprises you by lifting itself away from the table, without your permission or guidance. Who did that?

Your reflexes—there was no need for you to make any decisions.

Simple reflexes collaborate to produce complex behaviors. In the knee-jerk example, the extension of your leg requires more than the activation of one muscle group, like your quadriceps. The reflex circuit also sends signals that simultaneously inhibit other muscles, such as your hamstrings, which would normally oppose your leg flying forward, as their usual job is to produce flexion at the knee. The technical term for this kind of coordination between reflexes is *reciprocal inhibition*, which I like because it highlights the synergy and cooperation that occur between reflexive stimulation and inhibition. Remember how fight-or-flight and rest-and-digest work together? It's the same idea. Reciprocal inhibition in reflexes enables the nervous system to integrate simple signals to produce complex behaviors that are balanced, fast, and coordinated.

Let's say you've been reading for a while and lunchtime is approaching. You weren't aware as your body digested your breakfast, but now the emptiness of your stomach arrives as a sensation in your consciousness, prompting you to do something about it. Something like, say, pizza. Let's explore the diverse vagus nerve reflexes that collaborate to balance your organ functions during the simple act of eating.

Whether your preferred style is New York, New Haven, Chicago, or Neapolitan, or simply whatever is closest, eating a slice (or three) requires complex reflex cooperation. Upon entering the pizza parlor, your eyes scan the glistening options, a master class of neural analysis, where visual cues pair with memories and preferences. As you consider the toppings, your brain silently speaks, guiding you to your favorites. Slice in hand, your musculoskeletal reflexes and spinal cord, attuned to the position of your limbs in space, orchestrate the precise coordination needed to lift the pizza to your mouth. But even before those first movements, you are already salivating because your vagus nerve reflexes activated your salivary glands to lubricate your throat and prime your mouth for tasting, masticating, and swallowing. Meanwhile, other vagus nerve reflex signals to your stomach and pancreas increase the production of gastric acid and digestive enzymes. Systems at the ready, you take a bite and start chewing.

This cascade of reflexes is constantly attuned and adjusting to the rhythm of sensory feedback—the interplay of sensations from cheese, sauce, and crust on your tongue, the warmth of pizza traversing your pharynx and esophagus as it passes into your awaiting stomach. Such reflexive harmony between your digestive glands, tongue, and cheek muscles can even be stimulated by simply *thinking* about pizza. There is a very good chance you are salivating just reading this because even the act of consuming this story on the page can activate your vagus nerve reflexes.

As you eat, thousands of sensory neurons traveling up the vagus nerve from your esophagus, stomach, and pancreas deliver input to your brain, triggering thousands of reflexes. These reflexes stir your mind with thoughts and emotions, tapping into your memories of other pizza feasts and their pleasure. Meanwhile, these vagus nerve inputs to your brain stem stimulate reflex outputs that travel back down the vagus nerve to propel the food through your intestines, stimulate your pancreas to produce digestive enzymes and insulin, and synchronize this work with your heart and lungs.

You don't have to think about what your organs are doing with the pizza. You can just sit back and enjoy it, perhaps while you continue reading, or maybe you take a break and chat with someone. Pizza has fueled many a lab meeting.

Your reflexes never sleep, which is a very good thing. As you talk and chew, should a stray mouthful go "down the wrong pipe," your coughing reflex will be activated, expelling the errant morsels from your windpipe. And perhaps in all the excitement, you lose track and eat more slices than your normal one or two—as a doctor, I recommend you don't make it a habit, but who hasn't done it? Your bloating and discomfort will pass because your vagus nerve reflexes are hard at work, 24-7, collaborating to bring you back to homeostasis. Your food will move through the system, it will be digested, its nutrients will be absorbed, and the satiety signals to your brain will suppress your urges to keep eating. Your vagus nerve reflexes are always collaborating like this to balance your systems.

Reflexes are the key to understanding how the nervous system seamlessly orchestrates complex behaviors without conscious direction. By automatically integrating simple reflexes, the nervous system empowers us to perform complex feats of multitasking as varied as walking while chewing gum, playing a Mozart violin concerto while your heartbeat keeps time, or eating pizza and reading. Life and health depend upon the performance of unseen reflexes, protective and healing, that animate our existence and well-being.

These are but a few examples of the thousands of vagus nerve reflexes mediated by the vagus nerve fibers running up and down each side of your neck. Every one of these vagus fibers is unique, hardwired into its own specific reflex circuit. The great nerve's complexity can boggle the mind.

HEALTH DEPENDS ON HOMEOSTASIS

About 80 percent of your vagus nerve fibers are *afferent*, meaning they are sensory nerves transmitting input signals from your body's internal and external environment up to your brain. Each one of these fibers contributes specific information in the input arc of a specific reflex; so, each has a crucial role in mediating countless simple reflexes that together coordinate the healthy functioning of all your body's organs and systems while you are blissfully unaware.

A diverse range of chemical and mechanical stimuli activate vagus nerve reflex responses to changes in temperature, oxidative stress, and energy demands, and the presence of specific molecules like glucose, fat, nutrients, hormones, and cytokines made by the immune system. Your brain stem integrates thousands of such inputs and responds by activating the motor arcs in thousands of reflex circuits. These outputs, neural signals traveling back to the body via the vagus nerve, coordinate the overall behavior of your body's organs and vital systems. Each of your individual vagus nerve fibers plays its own part, in concert with countless other reflexes, to maintain your body in a healthy balance—a state known as *homeostasis*.

Homeostasis is the dynamic process through which your body, and the bodies of all living organisms, maintain a stable internal environment despite fluctuations both inside and outside of us. This regulation is essential for the proper functioning of your organs and vital systems and for your overall health, ensuring that conditions within your body remain within a narrow, optimal range.

Homeostasis is not encoded in your DNA. There is no gene for it. It is an emergent property, which is when an ensemble of simpler entities creates something unexpected and extraordinary, like a flock of birds winging across the sky in formation, or individual water molecules self-arranging into infinitely different shapes of snowflakes. Homeostasis is the balanced functioning of your physiological systems that emerges from the complex entanglement of hundreds of billions of neurons and trillions of synapses in your nervous system. Simple reflexes, selected by evolution, act together to produce stable organ systems, despite changes in the weather and food supply, despite injury or infection. Homeostasis emerges as something greater than the sum of its parts, something necessary to health, something that enables us to heal.

---

heal | hēl | *verb* to become sound or healthy again • alleviate (a person's distress or anguish) • correct or put right (an undesirable situation)

Billions of impulses, or action potentials, travel up and down your vagus nerve between your organs and your brain, participating in reflexes that produce homeostasis, facilitate healing, and sustain health. Reflexes maintain the seamless, rhythmic, balanced output of your cardiovascular, respiratory, gastrointestinal, hepatorenal (liver and kidneys), and nervous systems. This short list of vagus nerve reflexes represents only a few examples out of the hundreds that physiologists and neuroscientists have been studying for decades.

The baroreceptor reflex maintains blood pressure. This reflex starts with baroreceptors, specialized neurons that detect stretch, located in your carotid artery and in your aorta (the major artery near your heart). Their role is to continuously monitor changes in arterial blood pressure (oxygenated blood coming from your heart) from the stretch of vessel walls. When baroreceptors detect increased pressure because the walls are stretched, they send signals through your vagus nerve to the brain stem, which in turn activates vagus nerve motor fibers, increasing vagal tone to your heart and reducing your heart rate. Simultaneously, baroreceptors send signals to the brain stem neurons that dampen the sympathetic (fight-or-flight) nervous system, leading to blood vessel dilation and a decrease in blood pressure. Conversely, a decrease in blood pressure results in reduced baroreceptor firing, diminishing vagus nerve tone and enhancing sympathetic activity, which elevates your heart rate and contractility and causes your blood vessels to constrict to balance your blood pressure levels within a healthy range.

The Hering-Breuer reflex prevents your lungs from overinflating. When you inhale deeply, stretch-receptor neurons located in the smooth muscles of your lungs' airways (the bronchi and bronchioles) detect your degree of lung inflation. Impulses from these receptors travel up the vagus nerve to your brain stem. When the brain stem receives this input, your respiratory center located there sends signals that inhibit the inspiratory muscles, effectively terminating inhalation and initiating exhalation.

This prevents overinflation of your lungs and helps regulate the depth and rhythm of your breathing. When you practice deep breathing exercises, you stimulate this vagus reflex. We'll discuss this more in part 3.

Respiratory sinus arrhythmia (RSA) controls the natural variation in your heart rate that occurs during your breathing cycle. When you inhale, you activate pulmonary stretch receptors. Input signals from these receptors travel via the vagus nerve to the cardiorespiratory centers in the brain stem, which in turn send signals that inhibit vagus nerve motor output to the heart. The reduced vagal output increases your heart rate. This heart rate increase during inhalation, and subsequent decrease during exhalation, enhances the efficiency of the cardiac cycle by synchronizing heart rate with the breathing cycle, leading to optimal filling and emptying of the heart and optimal oxygenation of the blood.

The vaso-vagal reflex balances the propulsion, digestion, and absorption of nutrients during your every meal. Receptors located in the walls of the stomach and intestines are activated by the distension of the stomach or the presence of certain chemical stimuli, such as fats, proteins, carbohydrates, and gastrointestinal hormones. These signals are transmitted to the brain stem through the vagus nerve. In turn, the brain stem sends signals that control various digestive functions.

The hepatic glucose sensing reflex regulates your blood glucose levels. Vagus neurons sensitive to glucose levels in your liver modulate the output of vagus nerve motor neurons and sympathetic neurons that control glucose production and storage in your liver. Simultaneously, vagus and sympathetic nerves regulate insulin release from the pancreas, helping to maintain your blood glucose levels within a healthy range. This is one of the reflexes involved in the ultrasound therapy we'll investigate more in chapter 7.

The diving reflex optimizes your body's use of oxygen. This one is an internet and social media favorite. Cold water on your face stimulates sensory nerves, which send signals as input to your brain stem. The brain stem then coordinates an autonomic response, sending signals through the vagus nerve to your heart and respiratory system to decrease your heart rate and suppress your breathing. Additionally, the diving reflex redirects blood flow toward vital organs. This helps protect your body during submersion in cold water by balancing blood flow and conserving oxygen, which is why some people, especially babies, have been able to survive prolonged periods of cold water submersion and accidental near-drownings unscathed. As we will return to in part 3, cold exposure can activate this reflex.

Homeostasis is key to health, and the vagus nerve is key to homeostasis. Two centuries of research have revealed the vagus nerve's crucial role in regulating and balancing essential bodily functions. Your vagus nerve is working around the clock, harmonizing your vital systems to keep you well. Disruption of the vagus nerve's protective and healing reflexes upsets this balance.

A state of impaired homeostasis, called *dysregulation*, *disequilibrium*, or *disease*, happens when the body's internal environment is out of balance. Although not every instance of dysregulation is as extreme as what happened to Janice, persistent or severe imbalance in the reflexes controlling the body's organs and vital systems contributes to most, if not all, pathological conditions and states of adverse health. That is why understanding the reasons for vagus nerve reflex imbalances, and how to correct them, has emerged as one of the most critical pursuits in neuroscience, immunology, and all of medicine.

I am certain that what we know today about vagus nerve reflexes and health and illness is only the tip of the iceberg. As we discussed in chapter 2, studies at Harvard Medical School have demonstrated that as few as only one hundred vagus nerve fibers, representing less than 2 percent of all the vagus nerve fibers in a mouse, are enough to control the fundamentals of breathing. The specific functions of the vast number of vagus nerve fibers remain unknown. We don't know exactly how many fibers are required to control the input and output for most of the vagus nerve's reflexes, but it is likely to be a similarly small number. This suggests that by targeting specific vagus neurons with optogenetics and other tools, we can uncover new reflexes that control unexplored aspects of physiology.

Twenty-six years ago, my colleagues and I discovered a fundamental vagus nerve reflex that regulates the immune system—the very system that was long believed to have nothing to do with the nervous system. I'll explain more in chapter 5, but how we got to that experiment, how we discovered the inflammatory reflex, goes back to little Janice.

INFLAMMATION: WHEN IT'S TOO MUCH OF A GOOD THING

In the wake of Janice's death, I questioned whether molecules made by her own immune system, rather than bacteria, were the direct cause of shock and tissue injury. This possibility challenged the prevailing dogma that the immune system could do no harm and was only ever a force for good. But in the absence of any other explanation for her death, it seemed worth investigating. If true, then we could try to develop antidotes or antagonistic molecules to block the shock-inducing effects of the immune system, wielding these antidotes as new drugs to inhibit inflammation.

In 1985, heeding my grandfather's encouragement from childhood, I decided to do something about this, embarking on a research quest that continues to this day. Despite training to be a neurosurgeon, I had become fascinated by the molecular underpinnings of shock and inflammation—not because I knew the vagus nerve and inflammation were linked, not yet, but because my patient died in my arms. I did not know why, and I wanted an antidote to prevent others from suffering as she did.

Galen, with the keen eye of a poetic observer, recognized the essence of inflammation as the heat, redness, swelling, and pain that is the body's response to an injury or infection. The surprising truth about inflammation is that it's as intrinsic to our being as the cadence of our heartbeat, the ebb and flow of our breath, or the electric and chemical conversations whispering between our neurons. Inflammation, like all reflexes, is protective, our body's way of remaining whole and keeping going in the face of life's slings and arrows and microbes. Without our immune system and inflammation, the physician and author Lewis Thomas suggested, we wouldn't have selves: "Inflammation and immunology must indeed be powerfully designed to keep us apart; without such mechanisms, involving considerable effort, we might have developed as a kind of flowing syncytium over the earth, without the morphogenesis of even a flower." Because injury and infection are unavoidable consequences of being alive, inflammation is a universal reaction in the animal kingdom, the body's way of mending injuries and orchestrating the multifaceted immunological and metabolic defense strategies essential to ward off bacteria, viruses, and other lurking enemies.

Anyone who has lived with a compromised immune system, whether it's a result of a rare genetic disorder or, like Kelly Owens in chapter 1, from the influence of potent immunosuppressive substances meant to help them, knows how the body's ability to recover and heal is thrown into disarray. Wounds linger, and the onset of even minor infection grows threatening. Ultimately, your vagus nerve's ability to regulate inflammation and restore homeostasis during injury or infection may be the difference between existing or not.

So, the next time you encounter the pain of an infected tooth or a stubbed toe, try to find some solace, knowing that the pain is part of your protective inflammatory responses, a sign you are beginning the process of healing. As the white blood cells charge into the afflicted region, they stimulate your nerves, which hurts. But they are there to clear out debris, vanquish unseen invaders, and reweave your tissues. Remember this, too, when you are coming down with a cold or the flu because the onset of fever, loss of appetite, and fatigue mean your immune system is responding to the viral invaders. Specialized cells, called *phagocytes*, summon other immune cell defenders, called *lymphocytes*, which make lasting memories of the attack. The next time a similar germ invades, your immune system will rise in instant recognition, better equipped to ward off an illness. Inflammation is a testament to the resilience of life, where our cuts can mend, our immune system can remember, and our toes dance again.

If inflammation normally helps us in our daily fight to survive and prosper, how did it get its bad-guy reputation today? My medical school professors had stressed the benefits of inflammation to the patient. None had lectured about the possibility there could be too much inflammation, certainly not enough to kill. But this fable would change after I graduated and began my residency training in neurosurgery. It was Aesop, the ancient storyteller, who said it is possible to have too much of a good thing.

Eventually, my research would show that when inflammation is overstimulated, it can spin out of control and wreak havoc on the body's own healthy cells instead of mending the wound or warding off invading microbes. This discovery would lead me to more clues that the immune system, like the other organ systems, is controlled by the autonomic nervous system and that the vagus nerve has a central role to play.

In the early 1980s, I was training to be a neurosurgeon and living and working on Manhattan's Upper East Side at the intersection of three hallowed medical domains: the New York Hospital—Cornell University Medical Center (now Weill Cornell Medicine), Rockefeller University, and Memorial Sloan Kettering Cancer Center. Each one of these research institutions had a part to play in the unveiling of an important molecule made by the immune system that began to shed light on how inflammation is regulated.

After seeing what had happened to baby Janice, I wondered, Is it possible that molecules made by your body's own immune system can kill you? Could these same molecules be responsible for chronic inflammation conditions, albeit in lesser amounts over longer time periods? At this time, there was research being done on a molecule called *tumor necrosis factor*, or simply TNF. What little was known about its features suggested to me that it could be involved in lethal shock and tissue injury.

TNF was first identified in studies involving mice at Memorial Sloan Kettering. Researchers were administering TNF to mice in an attempt to eradicate cancerous tumors, hopefully without causing toxicity to the mouse. TNF is a cytokine, and cytokines are proteins, products of the immune system essential to inflammation. TNF, for example, signals the mobilization of other cells that kill invaders and initiate healing. It also prompts your body's metabolism to mobilize energy stores for these infection-fighting and recovery processes.

At this time, my colleagues and I were also studying TNF at Cornell in collaboration with Rockefeller University. But we were looking at its ability to cause inflammation nonspecifically by damaging normal tissues, not tumors. We discovered that more is not necessarily better when it comes to the immune system. And that since more can be too much, we had a first glimpse into the future possibilities for getting rid of TNF using monoclonal anti-TNF antibodies, now used by millions of patients.

The first time I infused TNF into a rat, I was alone in the early morning quiet of the lab. It was a surgical procedure similar to the one you witnessed in the last chapter: anesthesia, incision, separating the muscle layers to expose the carotid artery . . . In this case, after gently sliding a tiny polyethylene tube into the millimeter-wide artery, I connected the pulsing beat of the small creature's heart to a pressure transducer, which traced each beat onto a paper printout, like a ticker tape giving me written evidence of the rat's homeostasis scribed as blood pressure and heart and respiratory rates.

When I saw the transformation, the slide into shock, the deranged acceleration and then slowing of the rat's heart rate, and the cessation of breath, I knew I was seeing something for the first time: A molecule made by the body's own cells was doing the same lethal things to the body as the bacterial toxin lipopolysaccharide. The body was capable of killing itself with TNF.

In the necropsies, we found that the rat's kidneys were pale, the intestines blackened with necrosis, and the lungs swollen with fluid as if the animals had drowned. Further histochemical analysis on tissues prepared from these organs showed a microscopic parade of white blood cells because phagocytes, including macrophages and neutrophils, had invaded the tissues. The blood vessels in the lungs were so clogged with white blood cells that there was no room for blood to pass, a finding that caused or at least contributed to the onset of apnea and death.

Our data meant that every one of us is capable of inflammation excess. This is what had happened to Janice. Her body had made too much TNF, and it killed her. The realization was a powerful motive for us to move on to the next critical questions:

If TNF was the problem, could we use a TNF-blocking molecule to prevent inflammation? In other words, could we make an antidote to excessive inflammation? We understood that everyone's immune system can make TNF, yet dying from sudden shock and tissue injury is unusual. It was becoming clear that there must be a mechanism to regulate how much of the molecule the immune system makes to maintain health and homeostasis. I became obsessed with understanding it.

In the autumn of 1986, we embarked on the pivotal experiments that aimed to bring our concepts closer to human application. We crafted an antidote against TNF, an "anti-TNF monoclonal antibody," and we tested it in baboons.

Baboons are nonendangered primates found predominantly across various regions of Africa and the Arabian Peninsula that share many anatomical and physiological features with humans. Adult males weigh between sixty-five and ninety pounds. In many urban areas of Africa, baboons are seen as pests, much like raccoons in other parts of the world, boldly stealing food, raiding homes, and snatching suitcases from unsuspecting tourists. Although much more expensive and difficult to work with compared to mice and rats, baboons are a gold standard laboratory model to mimic the complex physiology of shock and tissue injury caused by lethal infections or inflammation in humans. With funding supplied by grants and contracts, we were able to get started.

Amid the hum of ventilators, fluid pumps, and hemodynamic monitoring machines, two tranquilized baboons rested supine on parallel operating tables in an animal surgery suite we commandeered on Cornell University Medical School's eighth floor. To replicate the clinical (human) scenario in every possible way, I ransacked a chest freezer filled with bacterial samples from shock patients treated over the years. Near the bottom, in a back corner that required me to climb nearly all the way into the freezer to reach it, I located a box containing a tube of bacteria that had killed a young man twenty years prior. The label identified the bacterial species as *E. coli,* a common cause of lethal infection in hospitalized, critically ill patients.

Using this living bacteria, we ignited the baboons' immune system to release TNF. Instead of using lab-made TNF, as we had administered to the rats, we wanted the infected baboon to find its own rhythm after it made its own TNF, as would occur in a patient. Then, we would try to save it from itself.

It turned out that we could.

On a clear autumn day in Manhattan, we began our side-by-side experiment in a windowless veterinary operating room suite. We infused one baboon with monoclonal anti-TNF, while the other animal, the control, received an irrelevant antibody that would not prevent TNF's effects. Then we perfused the veins of both animals with trillions of lethal *E. coli.*

For the next eight hours, we fine-tuned the administration of intravenous fluids based on their cardiovascular and hydration parameters using precise techniques mirroring the care you or I would get in a hospital intensive care unit. And as we did routinely with our human patients, we established protocols to record the clinical data every fifteen minutes, noting progress in flow charts as we watched and waited. What happened then changed my world. And in a few short years, it would change the world for millions of others too.

The reactions of the baboons to the bacteria in their veins were both revealing and deeply moving. Moments after bacteria had invaded their blood, TNF surged from within their immune systems, reaching its peak serum levels in just under two hours. One baboon, devoid of the anti-TNF monoclonals, was quickly overwhelmed, and like the rats, its blood pressure was soon plummeting and its life rhythms fading. By the eighth hour, its heart was completely still, and it succumbed. Yet its counterpart, with anti-TNF monoclonal antibodies flooding its bloodstream to mute the actions of TNF, was unaffected despite the bacteria replicating in its blood. On it continued with normal heartbeat, breathing, and kidney functions, undisturbed in homeostatic balance, hours after the time when its kin had died.

Although both baboons had been introduced to the same bacterial threat, the outcomes were vastly different. The protection of anti-TNF monoclonal antibodies was the difference between homeostatic harmony and systems collapse. This meant that the necessary factor causing lethal shock was not the bacteria but TNF. The baboon that lived because it received monoclonal anti-TNF proved this.

The day the rat died after I gave it TNF was a "holy-shit moment" of a magnitude scientists can expect to have just a few times in their lives, if they're lucky. The day the baboon lived because I gave it anti-TNF was another. It seemed certain to me and my colleagues involved in these first experiments, and then only to us, that this discovery about the immune system causing inflammation would disrupt medicine, and the pharmaceutical industry.

We stayed up all night. As dawn painted the sky, our antibody-protected subject revealed no signs of disrupted homeostasis. Then, as we had done so often before on human subjects, we gently removed the catheters that had connected our baboon patients to our world of inquiry and monitors. Once safely ensconced back in its home room, the baboon awoke, gazed sleepily at us, and began eating an apple as if touched by grace. It was a new day.

In December of 1987, we published these results in *Nature*, in an article demonstrating for the first time that monoclonal antibodies against TNF could be used as an experimental anti-inflammatory drug. This flew in the face of a major, time-honored premise in the field of immunology: that immune responses are, almost by definition, beneficial to the host. The stage was set for testing monoclonal anti-TNF antibodies in patients, an approach that would later become a new class of anti-inflammatory drugs now widely used by millions to treat diseases like rheumatoid arthritis and inflammatory bowel disease.

THE INFLAMMATORY REFLEX AND THE VAGUS NERVE

In 1992, I started a new lab at the Feinstein Institutes, in Manhasset, New York. For several years my colleagues and I studied an anti-inflammatory molecule we invented and named CNI-1493. It proved to be exceptionally effective in a wide range of laboratory models of harmful inflammation, including arthritis, shock, stroke, brain injury, and inflammatory bowel disease. It also worked in people: Clinical trial results showed that administering CNI-1493 decreased inflammation and tissue injury in patients with inflammatory bowel disease, and it decreased the production of TNF in patients being treated for cancer.

We had anticipated these results. What we did not expect was that the presence of CNI-1493 in the brain prevented the production of TNF not only in the brain but also in the body—a discovery we made thanks to an error in communication with the research team performing one experiment. Instead of giving the bacterial toxin (lipopolysaccharide) directly into the brain to observe the effects on TNF production being blocked by CNI-1493 in the brain, they injected the toxin into the mouse's abdomen. This should have normally stimulated the immune system to make high levels of TNF in the blood. When the results of serum measurements were in, we observed that animals with CNI-1493 in their brain had failed to produce TNF in their bloodstream.

How was the TNF blocker in the brain—in a dose too low to enter the bloodstream—affecting the body? There was no textbook explanation for this, so we investigated other possibilities. Inspired by explorers before me who had charted the reach of the vagus nerve, we found our answer. When we cut the vagus nerve and repeated the experiment, CNI-1493 in the brain failed to stop TNF production in the body. In other words, signals conveyed by the vagus nerve can stop TNF production.

I realized almost immediately the importance of this discovery. If the vagus nerve regulates the immune system through an anti-inflammatory, healing reflex, then it should be possible to regulate inflammation with devices that stimulate the vagus nerve in patients. And I was confident this could happen because, as a neurosurgeon, I knew that vagus nerve stimulation was even then already being routinely and safely performed to treat patients with epilepsy.

Ever since the Weber brothers slowed the frog heart and Otto Loewi discovered *vagusstoff*, we have known that stimulating nerves with bursts of electricity causes effects on the organs connected to the nerves. Armed with this simple, old-new idea, I wanted to see if stimulating the vagus nerve of a rat alone, without administering CNI-1493 or any other anti-inflammatory molecule, would reduce the amount of TNF.

To do the experiment, I first "borrowed" a small, handheld nerve stimulator, powered by two AA batteries, from the North Shore University Hospital's neurosurgery suite (a short walk across the campus from my lab). Armed with this simple tool, we electrically stimulated a rat's vagus nerve for five minutes before injecting the animal with lipopolysaccharide to stimulate the immune system into making TNF. Two hours later, we collected blood samples to measure the TNF amounts. In a few days, the results were in: Vagus nerve stimulation had suppressed the TNF levels by 75 percent as compared to control subjects. Only later did I learn that several members of a posse of scientists and technicians had bet against these results, but since they had not placed their bets with me personally, I do not know what our success cost them.

I do know that this was the third time I experienced a first-ever look at something no one else had seen. We had discovered a vagus nerve circuit that inhibited TNF, which we could turn on with a nerve stimulator. Holy shit, again.

There is a huge amount of literature about the relationship between the brain and the immune system, because for centuries doctors, practitioners of traditional medicine, and spiritual leaders of all stripes have talked and written about the mind-body connection and its role in health and illness. It seemed likely that the immune system, as our gatekeeper and frontline defense, might be a recipient or target of a mind or brain influence, but no one understood the mechanisms by which this could occur, and the limited scientific theories about these relationships were mired in philosophy and psychology, not neuroscience and immunology. For this reason, many mainstream scientists dismissed ideas about mind-body or nervous system–immune system connections as being of questionable validity—and it's why, for so long, they were not part of the medical school curriculum.

Robert A. Good, a founder of modern immunology, expressed his frustration this way in 1981:

> *Immunologists are often asked whether the state of mind can influence the body's defenses. Can positive attitude, a constructive frame of mind, grief, depression, or anxiety alter ability to resist infections, allergies, autoimmunities, or even cancer? Such questions leave me with a feeling of inadequacy because I know deep down that such influences exist, but I am unable to tell how they work, nor can I in any scientific way prescribe how to harness these influences, predict or control them. Thus, they cannot usually be addressed in scientific perspective. In the face of this inadequacy, most immunologists are naturally uneasy and usually plead not to be bothered with such things.*

Now they can be addressed in scientific perspective. They can, and we are. In my weekly lab meeting after that first vagus nerve stimulation had significantly suppressed inflammation, I sketched the connections between the brain, the vagus nerve, and our bodily organs like the spleen and liver, and had a realization that pulled all our findings together: the inflammatory reflex. I proposed that the vagus nerve carries an anti-inflammatory, healing reflex to add to the list of its previously known protective reflexes.

Inherent to its defensive duties, your immune system can also *sense* because its cells detect the presence of microbes and injured tissues. When this happens, the cells release molecules that stimulate your sensory neurons, causing the pain, fever, loss of appetite, and other sickness sensations. Neurons, we learned, sing back, not only to their neuronal kin but also to those cellular warriors of inflammation, macrophages and lymphocytes. As the immune cells sound their alerts using cytokines and myriad other inflammatory molecules, the neurons listen with attention, and sing their reply.

Beyond what you see, touch, taste, hear, and smell, your immune system, too, alerts your brain about your environment. This ensures you're always informed and responsive, reflexively, as necessary for your health and happiness. Neurons in your vagus nerve—your sixth sense—provide sensory arcs for your immune system's reflexes. This is the input. Recall, eighty percent of your vagus nerve fibers are sensory, capable of relaying prodigious quantities of information about the minute-to-minute status of your immune system! When the brain receives this information in the brain stem, where the vagus connects, it automatically directs the immune system to act accordingly through the motor arc of your immune system's anti-inflammatory, healing reflexes. This is the output. Input and output, all through the vagus nerve, sensing and responding, monitoring, adjusting, and readjusting to return you to homeostasis and to regulate inflammation. Your immune and nervous systems, rather than being siloed and distant, communicate in a shared electrical and chemical language.

Or, they do when everything is working as it should be.

If your vagus nerve isn't doing what it's supposed to, inflammation can run amok and overwhelm you. Then the inflammatory reflex presents an opportunity because we can reactivate and harness it with vagus nerve stimulators to inhibit inflammation. It's what your vagus nerve is made to do, but in some cases, it may require some assistance. In a world reliant on drugs to treat inflammation, this realization leads us to a new strategy—an entirely new field of medicine, offering vagus nerve stimulating devices instead of drugs (even anti-TNF) to patients who are not improving, or are afraid of the side effects, or cannot afford them.

The three experiments I've described in this chapter have had enormous implications. They explain that Janice's death was a result of rampant inflammation. Her body was unable to find its way back to homeostasis, an extreme failure of the inflammatory reflex. Today, what happened to Janice is more preventable. It is also rare. Much more common are chronic conditions like epilepsy, rheumatoid arthritis, inflammatory bowel disease, diabetes and obesity, and depression and anxiety. As science and innovation converge, the vagus nerve emerges as a therapeutic target that can prevent or reverse the inflammation behind these conditions. And learning how to stimulate the vagus nerve brings us closer to a world where catastrophic inflammation can be averted, where inflammatory diseases that affect millions are better treated.

How do we do this? And does it work? These are the questions of part 2.

II

Great Interventions

The New Frontier of Stimulation Therapy

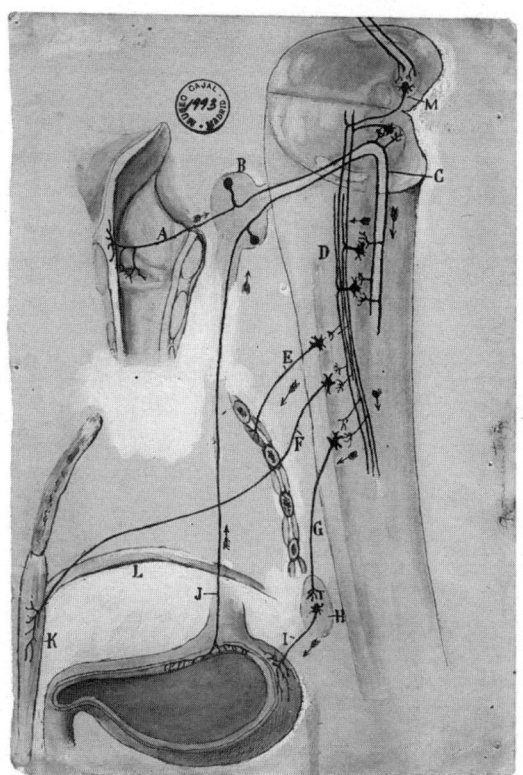

4

The Path to Stimulation and Early Experiments with Epilepsy

Life and death appeared to me ideal bounds, which I should first break through, and pour a torrent of light into our dark world.

— MARY WOLLSTONECRAFT SHELLEY, *FRANKENSTEIN*

As anyone who has seen a tree charred by a lightning strike can attest, meetings between electricity and living cells are fraught with potential danger. Electricity is a flow of energy produced by subatomic particles called *electrons*, whose movement through the biological fabric of living tissues can generate enough heat to destroy a tree trunk or the cells in our body. The amount of heat released depends on the force driving the electricity (defined in volts) and the resistance to the flow of electricity (defined in ohms). In the United States, household circuits operate at 120 volts, and children learn from an early age not to put forks or fingers into electrical sockets. A single lightning flash can release over three hundred *million* volts, capable of producing enough heat to ignite a tree or kill an unfortunate golfer.

So, electricity might seem like an unlikely therapy. How is it possible to harness it for long-term use in a vagus nerve stimulating device without destroying the nerve? How can electricity be good for us? Why isn't "bioelectronic medicine" a contradiction in terms?

The answers lie in the fundamental difference between the flow of electric current and the flow of information signals in neurons that travel from one end of the nerve fiber to the other. Electricity requires a complete loop, called a *closed circuit*, for energy to move from one place to another. When you turn on a switch at home, you complete or "close" that loop, and power flows along the wire traveling to the light bulb or appliance. Turning off the switch breaks the loop and stops electrons flowing through the wire, interrupting the transfer of energy from one point to another in the electrical circuit. With lightning, the tree or the golfer is effectively the switch, closing the loop between the lightning and the ground. Electrons flowing against resistance in a wire generate heat in proportion to the resistance, which causes electrical energy to be converted into thermal energy. When resistance is high, or the voltage pushing the current is high, the amount of heat produced is also high. Dangerously high.

But signaling in neurons is different. Electrons don't flow freely along neurons as they do in a conducting wire. Rather, as I described in chapter 2, neurons propagate distinct spikes of electrical signals, known as *action potentials*, which move along the neurons. Each spike occurs because of a voltage difference across the cell membrane, called a *membrane potential*, caused by the differential distribution of charged molecules (the molecules bearing either negative or positive charges). Small pores, called *ion channels*, are strategically positioned along the length of the neuron's membrane. These allow charged molecules, called *ions*, to flow from inside to outside, and from outside to inside, of the neuron across the neuronal membrane. This flow of ions in and out of the neuron causes the voltage changes, or action potentials.

Action potentials travel the length of the neuron as the adjacent ion channels open in sequence, one after the other from the beginning to the end, like a row of dominoes knocking each other forward. The action potentials do travel fast, in biological terms, but their speed is quite slow compared to the flow of electricity in a wire. For example, the speed of electric current in a conductor is close to the speed of light, at nearly three hundred million meters per second, but the speed of action potentials moving along a neuron ranges from only 1 to 120 meters per second.

To harness electricity to safely stimulate the vagus nerve, we use rhythmic bursts of electricity that mimic action potentials rather than the continuous flow of electrons in lightning or electrical wiring. We also carefully sequence the delivery of each pulse to coax the nerve into firing its own action potentials without producing dangerous quantities of heat or overtaxing its delicate structures. We do this with an electrical pulse generator, an instrument housing an electrical circuit designed to rapidly toggle between a higher voltage level output and a lower voltage level output, like quickly turning on and off the electric current to the electrodes. The precise timing and sequence of these voltage changes create a characteristic "square" shaped wave.

These electrical impulses pluck the vagus nerve, producing action potentials like musical notes. The tempo of a vagus nerve stimulator's electrical output, known as frequency and denoted in hertz (Hz), decides the pacing of stimulation, or how many times the neurons are prompted to fire within a second. With music, we hear frequency as pitch. Pulse width, measurable in the currency of milliseconds, determines the duration of each individual note, affecting which specific neurons in the vagus nerve join in the concerto. Finally, amplitude, like a volume dial, adjusts the intensity of the electrical waves coursing through the electrode. These are the key adjustable parameters for any vagus nerve stimulating device: the strength or amplitude of the current, the width of each pulse, and the frequency of the pulses transmitted through a wire to a pair of electrodes. By tuning each of these parameters, we can generate a precise stimulation tailored to specific kinds of nerve fibers within the vagus nerve itself to select those strands that play key roles in treatment, like stopping inflammation or halting a seizure.

The physiological effects of vagus nerve stimulation on the brain and body begin when electrodes initiate neuronal firing, and then extend deeply into the nervous system, because individual action potentials pulsating in vagus nerve fibers are part of a vast, interconnected neural network. You'll recall Otto Loewi demonstrating that action potentials transmitting along the length of your vagus nerve incite the nerve endings to release neurotransmitters, or "vagus stuff," into the synapse. And remember that specific vagus fibers control specific aspects of physiology, like breathing in and breathing out. Today we know that there are billions of synapses in the nervous system that are activated when the vagus nerve is stimulated. Thus, by accessing the vagus nerve with an electrode in your neck, we are beginning to learn how to modulate the physiological harmony of your body and your brain.

You might be surprised to learn that hundreds of thousands of people today already live with a surgically embedded vagus nerve stimulator (VNS), a medical marvel rooted in nineteenth-century practices. This is how we know they are safe as we continue to test their effectiveness for treating different illnesses. Most VNS devices have been placed in patients who have epilepsy, the first disease to be treated by vagus nerve stimulation.

To be clear, the historic treatments for epilepsy using electricity were not very safe, very effective, or at all gentle. By today's standards they were ghastly. In a late-1800s innovation in the United States, patients in the care of epilepsy specialist Dr. J. L. Corning received "protracted applications" that combined carotid compression (of the carotid arteries, in the neck, that supply blood to the brain) with electrical stimulation. Corning invented "a special appliance, which, instead of being held in the hand, is secured by means of a band, which encircles the neck and is fastened behind by means of a buckle. . . . With this appliance it is possible to make applications lasting one or many hours. All that is necessary is a chloride-of-silver battery . . . and periodic moistening of the electrodes." He designed this device to stimulate the vagus nerve to slow the patient's heart rate, at the time believing that epilepsy was caused by excessive blood flow to the brain.

According to Dr. Corning, the effects of his "neck truss" were "most remarkable," but not without complications that included generalized weakness, overwhelming fear, physical agony, and emotional turmoil, manifesting through episodes of sobbing, vertigo, fainting, and occasionally death. (What happened to "first, do no harm"?) We can only wonder how anyone could endure such an ordeal. Maybe Dr. Corning's patients were sustained by their unshakable trust in the doctor, convinced he knew what he was doing. Certainly, his account makes me appreciate that today most of us place our trust in evidence.

Still, Corning's "neck truss" contained the seeds of ideas that became some amazing technology we have today. It was the predecessor of bioelectronic vagus nerve stimulation therapy for epilepsy, inflammation, depression, and other applications we'll explore in the following chapters.

Vagus nerve stimulation as a treatment for epilepsy has been around longer than you might think, which is not only interesting to me but also reassuring to anyone worried about "newfangled" devices and safety. The modern birth of vagus nerve stimulation began in the historic city of Pisa, Italy, in the early 1950s, a period of postwar transition, aspiration, and invention. In a laboratory at the Istituto di Fisiologia at the University of Pisa, Alberto Zanchetti was studying epilepsy in animals and meticulously documenting findings on electroencephalograms (EEGs), a method used to measure the brain's electrical activity and the presence of seizures: "The spindles disappeared abruptly as soon as [the vagus nerve] stimulation commenced," he wrote. "The interruption was definitely due to vagal impulses, because such influence was not observed if the proximal end of the vagus had been tied."

Zanchetti delivered electrical pulses lasting only 0.5 milliseconds, or half a thousandth of a second each, to a vagus nerve of an anesthetized cat. This incited the vagus nerve to transmit action potentials traveling up to the brain. A tiny pitchfork electrode had converted the power of electricity into natural vagus nerve signals, giving him the first glimpse at using a vagus nerve stimulator to quiet epileptic EEG spikes.

But for understandable reasons, Zanchetti's discovery failed to gain notice or generate excitement for clinical use. Until then, no one had shown it was even remotely possible to safely harness electricity to treat a clinical condition. And no one had ever safely implanted an electrical device into a human since all the available electrical pulse generators were bulky vacuum tube–powered machines designed for use in laboratories.

It would take the development of the now nearly ubiquitous technology, the heart pacemaker, to push medicine toward cutting-edge twenty-first-century vagus nerve stimulation. In the early 1950s, a researcher named Dr. Paul Zoll was doing parallel work to Zanchetti four thousand miles across the Atlantic in Boston. Dr. Zoll was looking for a new way to help patients with Stokes–Adams disease, a rare medical condition characterized by a disorder of the heart's electrical system that causes an irregular heartbeat and slowing of the heart, or bradycardia. The slowed heart can cause a sudden loss of consciousness, and sometimes, if the heart beats too slowly, it is fatal. In some Stokes–Adams patients, episodes of bradycardia can occur many times a day.

In Zoll's time, these near-death episodes were treated with a long hypodermic needle to inject drugs directly into the heart. Picture the violent scene from *Pulp Fiction*, when John Travolta's character plunges a syringe of epinephrine into the heart of Uma Thurman's character to resuscitate her from a heroin overdose. But now play that scene over and over again, as some patients would need this violent resuscitation several times a day, and treatment starts to look less like medical intervention and more like the famous shower scene from a famous horror movie. There had to be a better way.

Thanks to Zoll and his colleagues, now there is. In 1952, Zoll restarted a Stokes–Adams heart by applying electricity to the chest of a sixty-five-year-old man with severe coronary disease. A few years later, a heart surgeon in Minneapolis would take electrical stimulation technology further, implanting onto the heart of a three-year-old girl a Teflon-coated wire that could be connected to a bedside electrical pulse generator. This procedure has been replicated millions of times since and continues to happen daily in operating rooms around the world.

Although these events launched a new era of implanted electrical wires, the early electrodes were limited by the requirement that they be attached to an electrical pulse generator, which in those days resembled the bulky household radios of the era. Being hooked up to your pulse generator was like being tied to a major household appliance stationed next to your bed. Powered by large vacuum tubes, pulse generators were unwieldy, prone to breakdowns, and dependent on external power sources and extension cords.

Then came the transistor. Made from materials such as silicon or germanium, which are semiconductors (meaning they conduct electricity with efficiencies ranging between that of a conductor, like copper, and an insulator, like plastic), the conductivity of a transistor can be precisely controlled and modified, making it possible to amplify or switch electronic signals nimbly and exactly. Almost overnight, gone were the bulky furniture-sized devices. In their place, small transistor-powered devices were programmed to administer intermittent electrical pulses, carefully calibrated and timed, in sync with the doctor's prescription. Enabled by these tiny new technological building blocks, inventors harnessed computers within sophisticated medical instruments that were small enough to be implanted within the human body to modulate vital functions like the beating of your heart and the signals of your vagus nerve. And as technologies developed, the small devices kept getting smaller.

But the problem of external power sources remained. During a three-hour power failure in Minneapolis in 1957, a baby who was a patient of a doctor named C. Walton Lillehei died while recovering from surgery because the electricity supplying the pulse generator died. Angry and frustrated, Dr. Lillehei pleaded with a colleague to build a new, portable, battery-powered device. The colleague, Earl Bakken, was an electrical engineer, inventor, TV repairman, and founder of the medical device company Medtronic, who had been fascinated by electricity since he saw the movie *Frankenstein* as a nine-year-old boy. He agreed to try. He recalled reading an article in *Popular Electronics* that gave instructions for using transistors to make a battery-powered metronome that one could place on the piano. A few weeks later, Bakken gave Lillehei a sort of battery-powered metronome for the heart.

As he was passing through the hospital the following day, Bakken was stunned to chance upon a child fresh from open-heart surgery, already wearing his transistorized prototype. In his autobiography, Bakken recalls Lillehei's explanation for this lightning-fast adoption of the new device: "In his typical calm, measured, no-nonsense fashion he explained that he'd been told by the lab the pacemaker worked and he didn't want to waste another minute without it. He said he wouldn't allow a child to die because we hadn't used the best technology available."

In a mere four weeks, the first battery-powered, transistorized external pacemaker was launched for clinical use. And not long after, though clinical trials were still in progress, a surgeon in Stockholm, Sweden, implanted the first battery-powered, transistorized pacemaker, about the size of a hockey puck, in the abdominal wall of a beloved husband whose wife insisted. For better and for worse, these rapid achievements would be impossible to replicate in today's complex regulatory and ethical landscape, governing devices at all stages from their early development to clinical testing and use in patients.

Prior to 1957, the year I was born, entirely implantable electrical devices were nonexistent. As I write, millions of such devices are in use worldwide (and Bakken's company, Medtronic, enjoys annual sales exceeding $31 billion). More than three million battery-powered, transistorized, fully implantable cardiac pacemakers have been placed into patients around the world. Beyond those particular lifesaving devices, the development of cardiac pacemaker technology also established a platform to build a new kind of battery-powered, transistorized, computerized, fully implantable stimulation technologies targeting nerves, including the vagus.

EPILEPSY AND THE FIRST VNS PATIENT

The technological advances of cardiac pacemakers and epilepsy research finally converged into vagus nerve stimulation in the mid-1980s. In Philadelphia, not far from where Benjamin Franklin had once flown a kite to harness electricity from lightning, neuroscientist Jacob Zabara was working in his lab at Temple University. He was aware of Zanchetti's experiments in Pisa and realized that it should be possible to adapt cardiac pacemaker technology to do what his Italian predecessor could not have done without the technology of the intervening decades. Like Zanchetti, Zabara wanted to use vagus nerve stimulation to help people with epilepsy. He teamed up with an electrical engineer with career experience in designing pacemakers, and together they established the company Cyberonics, dedicated to advancing epilepsy treatment. Leveraging off-the-shelf integrated transistor circuits and receiving design assistance from colleagues in the field, Cyberonics developed an implantable device featuring a helix-shaped electrode to wrap around the vagus nerve in the neck.

In November 1988, a young man named Toney Kincaid became the first patient to receive this implantable vagus nerve stimulator for any condition. Toney had been afflicted with debilitating epileptic seizures daily for five years since developing an arteriovenous malformation, a tangle of big arteries and big veins in his brain. Mostly confined to his home, except for trips to the doctor, Toney was fortunate to find his way into a study of vagus nerve stimulation therapy for epilepsy.

The word *epilepsy* aptly finds its origins in the Greek word *epilambanein*, meaning "to seize." Indeed, it's a condition that can seize individuals suddenly and without warning. In its early history, epilepsy was often shrouded in a veil of superstition, regarded as a form of divine possession that both elevated and alienated those affected. Yet, amid these mystical interpretations, the ancient physician Hippocrates broke new ground. In his pivotal work "On the Sacred Disease," the author of the oath sworn by new physicians to this day proposed a then-revolutionary idea that epilepsy, just like other maladies, came from natural causes in the brain. This was a profound shift toward recognizing epilepsy as a medical disorder rather than a divine curse, and it set the stage for the treatments we use today.

We now understand that epilepsy's mechanism is rooted in brain neurons. As medically defined, epilepsy is a chronic neurological disorder marked by recurring seizures that originate from abnormal, synchronized neuronal firing in the brain. These electrical cellular events spread from neuron to neuron like a lightning storm.

Epileptic symptoms and signs can manifest subtly as brief episodes of unresponsiveness. In other cases, it can escalate into forceful convulsions and complete loss of consciousness, the "black darkness" blotting out everything that Dostoyevsky described when he drew on firsthand experience for the character of Prince Myshkin in *The Idiot*. Seizures can be localized to specific brain regions or generalized to affect larger brain areas. Some seizures produce loss of consciousness, muscle spasms and rigidity, and uncontrolled movements ranging from subtle to vigorous. In other cases, milder seizures may manifest as episodes of staring or subtle repetitive mouth and limb motions.

The clinical nature of a seizure depends upon the location in the brain of the specific neurons undergoing electrical upheaval. One person with epilepsy might experience visual hallucinations when epileptic discharges occur in the visual cortex; but another might have epileptic firing in the temporal lobes—a key brain region for emotions and memories—which triggers diverse sensations like déjà vu or sexual arousal. Sometimes temporal lobe epilepsy produces distortions of one's own bodily image, the sensation of Alice becoming very large or very small after falling down the rabbit hole, known as Alice in Wonderland syndrome. If you have epilepsy or know someone who does, you know it can be disruptive and hard to live with. In severe cases like Toney's, it is debilitating. People with epilepsy may be limited in everyday activities like driving and the kind of work they can do. The prospect of having a seizure in public understandably can cause anxiety. Friends and family adjust their lives to keep their loved one safe. And unfortunately, even in the twenty-first century, epilepsy is widely misunderstood and stigmatized. Even today, two thousand years after Hippocrates, some religious groups continue to practice rituals to excise demons they believe are causing seizures.

Beyond the clinical signs and symptoms of epilepsy, diagnostic procedures use EEGs to measure and record electrical activity in the brain. By affixing electrodes to the scalp, it is possible to measure the electrical patterns produced by activity in the brain's neurons. In cases of epilepsy, these EEG recordings frequently trace distinctive "spikes" in brain waves—characteristic electrical discharges in neurons indicative of epileptic activity. These EEG spikes, paired with corresponding clinical manifestations and with brain imaging techniques like MRI and CT scans, establish the diagnosis.

Upon confirming a diagnosis of epilepsy, first-line therapy is a class of drugs called *anti-seizure medications* or ASMs. With over twenty available ASMs, the selection process is guided by maximizing the benefit to prevent seizures while mitigating potential risks of side effects. The choice of seizure medication for a specific patient is not governed by a universal formula because no single medication is always effective, and each bears its own array of side effects, manifesting across a range of discomforts, including dizziness, drowsiness, fatigue, cognitive shifts, and emotional fluctuations. Each patient's treatment journey is unique. Through trial and error, doctor and patient find the optimal medication and dosage—hopefully. The journey can feel more like a maze.

Nearly one-third of epilepsy patients cross the threshold for needing surgical intervention. Neurosurgeons may open the skull to implant sheets of electrodes onto the brain surface to record neuron activity with a precision far exceeding EEG. These cortical recordings, collected over a period of several days while the hospitalized patient's every movement is also videotaped for later behavioral analysis, can pinpoint the origin of seizures in the brain. Suitably armed with brain imaging and cortical recordings, it is then possible for the neurosurgeon to identify a precise location of the seizure focus and excise the epilepsy epicenter.

Alternatively, neurosurgeons today may install an implantable electronic device into the neck to electrically stimulate the vagus nerve, thanks to the groundbreaking surgery performed on Toney more than thirty years ago. The study in 1988 involved twenty subjects, including Toney, in four different locations who would receive the implant. Toney, his neurosurgeon, and the research team at Wake Forest Bowman Gray School of Medicine in Winston-Salem, North Carolina, were all ready to go as soon as the FDA approved the study, which is how he became the first VNS patient.

My friend William O. Bell, or "Bill" to me, performed the surgery. He made an incision in the left side of Toney's neck, along the lower two-thirds of the anterior border of the sternocleidomastoid muscle. After separating the muscles and fascia, the left vagus nerve was isolated from its neighboring carotid artery and jugular vein. Then a stimulating electrode, which looks like a miniature Slinky, was carefully wrapped around the vagus nerve. (In surgeries like this, it is important to avoid the recurrent laryngeal nerve, the one Galen identified two thousand years ago as controlling the voice box in his squealing pig.)

Next a *passer*, as it's called—a temporary tube through which the surgeon threads the electrode wires—was tunneled under the skin to a pocket in the chest below the clavicle, or collarbone. After connecting one end of a wire to the pulse generator, which rested in the pocket, and the other to the vagus nerve electrode in the neck, the surgeon tested the device and closed the incisions. Toney experienced no generalized weakness, overwhelming fear, physical agony, or emotional turmoil; no sobbing, fainting, or vertigo, and he is alive and well today.

Prior to receiving his implanted vagus nerve stimulator, Toney had an average of eighty seizures a day. But after surgery, under the care of his neurologist, the strength of the device's stimulation was incrementally increased over a span of nine months, and the frequency of his seizures declined, from more than two thousand a month to only a few each day. On July 31, 1989, eight months after Toney's VNS device was implanted, a full day passed without a single seizure, his first seizure-free day in years.

Thankful and hoping to give back, Toney founded the Epilepsy Association of North Carolina, a 501(c)(3) that he currently leads as CEO to raise awareness about epilepsy and support research for new therapies. Their mission is "to speak for those who cannot, stand for those who aren't able, educate those hurting, bring education to the unlearned, give love and support to the despaired, be a light for those living in the shadows and be a friend to those who feel abandoned."

"There are no more grateful patients than the epilepsy patients you intervene on and make seizure-free," Bill, in retirement now, told me recently. "There's nothing else wrong with them, and you fix what's wrong, and they get to fully live life—get promoted to positions at work that they couldn't do before, drive again or get their driver's license for the first time."

In the initial study of the first four epilepsy patients treated with an implanted vagus nerve stimulating device, including Toney Kincaid, three out of the four experienced a significant reduction in seizures. Two saw a 100 percent reduction in seizures, one observed a 40 percent decrease, and the fourth had no improvement. Considering that all four patients had exhausted hope for therapy using anti-seizure medications, these results were indeed promising. The side effects were mild, including transient hoarseness (because of feedback via the recurrent laryngeal nerve) and a sensation of buzzing or muscle vibrations in the neck when the stimulation was activated. One patient experienced uncontrolled hiccups, but none of the patients reported significant bradycardia or other serious problems.

To make the implantable vagus nerve stimulator widely available in the 1990s, the small study with Toney Kincaid and a handful of others was repeated in carefully controlled clinical trials approved by the FDA. Building on the pilot studies, Cyberonics sponsored a larger, double-blind, randomized study at the University of Gothenburg, on the west coast of Sweden. After a twelve-week baseline, identical VNS devices were implanted into all subjects, but patients were randomized into two treatment groups based on the amount of electrical stimulation delivered—high or low. The findings from the first sixty-seven patients in this fourteen-week phase were striking. Those in the high VNS treatment group showed a mean seizure frequency reduction of 30.9 percent, a significant improvement over the 11.3 percent mean reduction observed in the low VNS group. Notably, close to 40 percent of patients in the high VNS group achieved at least a 50 percent reduction in seizure frequency, twice as many as in the low VNS group. Based on these results, the FDA approved the use of vagus nerve stimulation to treat epilepsy, which had previously only been treated with drugs.

Today, the implantation of a vagus nerve stimulating device is a relatively straightforward, typically uneventful procedure. The surgery, performed with either local or general anesthesia, is usually an outpatient procedure, meaning most patients go home the same day. Side effects are rare (affecting less than 5 percent of patients); the most common include minor infection, vocal cord weakness, neck pain, and exertional shortness of breath (called *dyspnea*) during the recovery period. These surgery-related complication rates are similar to the risks of other devices implanted in millions of patients, including pacemakers and artificial joints. These risks may be significantly less onerous than the potential consequences of uncontrolled seizures, including disability and ongoing exposure to other medication side effects.

With hundreds of thousands of vagus nerve stimulators now implanted for epilepsy, what have we learned about their effectiveness? Does it help people who have not found much relief from medications?

I am happy to say, in a word, yes. Not everyone, but many; not perfectly, but significantly. The scientific literature on this topic is complex and lengthy, but it reveals a range of responses to implantable vagus nerve stimulation, generally measured in the percentage of patients experiencing at least a 50 percent reduction in the frequency of seizures. These percentages range between 45 and 65 percent, yet the story does not end there. Children with severe epilepsy syndromes like Lennox-Gastaut, Landau-Kleffner, and Dravet have also seen benefits, with significant reductions in both seizure frequency and severity.

A question remains: *How* does vagus nerve stimulation work for epilepsy? This would not be the first or only instance of neuroscience knowing that something works without understanding the mechanisms. Years of research and clinical experience from hundreds of thousands of patients have led to a series of hypotheses, ranging from the dampening effect of vagus nerve stimulation on hyperactive brain regions to modulating brain neurotransmitter release. Some researchers have focused on increases in brain norepinephrine as a potential key, while other emerging evidence points to inflammation driving epileptic cascades. This suggests that VNS may offer benefit in epilepsy because it suppresses inflammation.

Some researchers propose that VNS therapy may sway the balance of certain neurotransmitters like gamma-aminobutyric acid (GABA) and glutamate, key players in the symphony of neural excitability. By modifying the delicate equilibrium between excitatory and inhibitory impulses, vagus nerve stimulation might avert the onset of seizures. It is also possible that vagus nerve stimulation triggers the release of endogenous opioids, the body's natural painkillers that may also have anti-seizure properties. Still another theory proposes that—ever so much more gently now than in the nineteenth century—vagus nerve stimulation influences cerebral blood flow, a subtle result that might indirectly modulate neuronal activity, potentially decreasing the intensity or frequency of seizures. Advocates of this theory point to a significant correlation between increased blood flow to the thalamus and a change in seizure frequency, but the scientific community has yet to reach a consensus on whether these changes are directly tied to vagus nerve stimulation's therapeutic effects.

Beyond blood flow, vagus nerve stimulation may influence the brain's very structure. Over time, VNS may foster neural plasticity, the ability of neurons to form new connections and reorganize existing ones. This could make the brain more resilient against seizures by strengthening and weakening some of its synapses.

Although each of these theories offers a possible explanation for why vagus nerve stimulation can reduce epileptic seizures, the short answer is we really don't know how it works. But the widespread use of these VNS devices also had direct and important implications for our ability to study how the vagus nerve controls inflammation.

5

Rebalancing Inflammation

I had come into the presence of a technological marvel, namely me.
—LEWIS THOMAS

Each year, sixty million human beings die, forty million—*two-thirds*—of them ensnared by their own out-of-balance immune system. Although their death certificates do not name *inflammation*, it's there, lurking behind the disease brands we're more familiar with: heart disease, stroke, Alzheimer's, diabetes, pneumonia, chronic obstructive pulmonary disease, cancer, and more, each one of these either caused by inflammation or made worse by it. Others suffer from autoimmune conditions like rheumatoid arthritis or inflammatory bowel disease and live with painful disabilities.

Back at the Feinstein Institutes, by the time I became interested in the relationship between the vagus nerve and inflammation, in the first decade of this century, vagus nerve stimulation therapy for epilepsy had become common. Each week, patients with epilepsy were having stimulators implanted at hospitals near my lab. Sangeeta Chavan and I seized on this opportunity to answer an important question: Does vagus nerve stimulation suppress TNF release by the human immune system? We conducted a study with seven epilepsy patients who were scheduled to receive a vagus nerve stimulator.

None of the patients we studied had an inflammatory or autoimmune disorder, but they consented to donate some blood for our research study designed to assess whether their VNS device would suppress their immune system's ability to produce TNF. We collected blood samples before, during, and after the vagus nerve stimulator was implanted, then back in the lab Sangeeta added endotoxin to the tubes of blood. This stimulated the white blood cells in the blood to begin making TNF, which she could then measure in the tubes. The results would lead us to another important breakthrough.

Sangeeta, a brilliant and enthusiastic scientist, shines when speaking about science. She learned her rapid-fire English in India before emigrating to the United States as a postdoctoral fellow in immunology. At the Feinstein Institutes, she was promoted up the academic ladder to full professor. Now, as cohead of the lab with me, she describes the scope of our work this way: "Everything cannot depend on one protein, in one cell type, right? It has to be the whole animal picture. So, we don't study one cell type or only one disease, because we are interested in what happens as a whole when different physiological systems are in play and we want to find cures."

After comparing the amount of TNF produced in the blood collected before activating the vagus nerve stimulator in epilepsy patients to how much was produced after activation, the answer was clear and unambiguous. Turning on the vagus nerve stimulator significantly decreased the amount of TNF the immune cells produced. When Sangeeta and I saw those TNF results, we were convinced that in the future, patients with inflammation would be treated using vagus nerve stimulation because the ability to control the amount of TNF production is an extraordinary gateway into treating a long list of inflammatory conditions.

For this to happen, I needed a company, so in 2005 I cofounded SetPoint Medical with my colleague and friend H. Shaw Warren. As an infectious disease specialist at Massachusetts General Hospital, Shaw had spent decades studying novel methods to treat inflammation. We dreamed up a new bioelectronic enterprise with a mission to conduct clinical trials on vagus nerve stimulation (VNS) to test whether my lab's discovery of the inflammatory reflex might work in people as well as it did in mice. Shaw and I had a long history of collaborating on research projects and launching new ventures, including cofounding Critical Therapeutics, a biotechnology firm to pioneer treatments for sepsis and other ICU-related conditions. We spoke daily, churning and polishing ideas and considering the ways that electronic devices could one day replace expensive, toxic drugs.

Shaw and I chose the name *SetPoint Medical* because, like the thermostat on your wall controls the temperature set point in your house, we intended to use the vagus nerve stimulator to establish a new healthy set point on the immune system. We thought SetPoint captured the notion of exerting functional influence over a complex system and regulating it for the better.

The FDA approval process for medical devices like vagus nerve stimulators typically begins with clinical trials designed to assess the safety of the device in a small number of patients. If the device is shown to be safe in these early trials, the manufacturer can then request approval to conduct clinical trials designed to assess the device's efficacy in a larger sample of people. These trials typically compare the new device to the standard of care treatment for the condition that the device is being used to treat. If the new device is shown to be more effective than the standard of care treatment, or if it is shown to be an equally effective treatment with fewer side effects or lower cost, then the FDA may approve the device for marketing.

These stages of FDA approval are sometimes described in three phases, where Phase 1 primarily assesses the safety and tolerability of the device or drug, Phase 2 evaluates its preliminary efficacy and continued safety, and Phase 3 confirms its efficacy on a broader population while monitoring adverse reactions, comparing it to standard or equivalent treatments. Beginning with a new idea from a laboratory, then starting the clinical translation process from scratch to move through all three phases to gain approval takes many years. And it costs many millions of dollars. The failure rate is high, which means that the risks to the investors and trial sponsors are also high. But eventually, if a new medical device is shown to be safe and effective in these clinical trials, the manufacturer can then submit an application to the FDA seeking approval to market the device.

Because rheumatoid arthritis and inflammatory bowel disease were our lead clinical targets, Shaw and I knew SetPoint would face serious challenges. As a cash-constrained medical device start-up company, we would be going head-to-head against existing multibillion-dollar pharmaceutical companies marketing anti-inflammatory drugs, including the anti-TNF antibodies I had helped bring into being with my publication in *Nature* twenty years earlier. Biological anti-inflammatory drugs related to monoclonal anti-TNF antibodies had become franchise products at these huge companies, generating billions of dollars of profit annually worldwide. Shaw and I knew that SetPoint's upcoming clinical trials would cost tens of millions of dollars, and that ultimately we would have to raise upward of $200 million to garner all of the necessary FDA approvals before the first VNS device to treat inflammation could be put on the market. In the beginning, though, our immediate goal was to convince some of the best and most talented people to come work at the company.

Yaakov Levine, an exceptional graduate student from my lab, became one of SetPoint's pioneering team members. Yaakov launched a research division within the confines of the Feinstein Institute, utilizing lab space that SetPoint leased. This arrangement facilitated seamless collaboration between the company and my lab, a mutually beneficial arrangement because we could assist Yaakov with any challenges related to the molecular biology or neuroscience of the inflammatory reflex, and he could assist us with the latest advances in biomedical engineering applied to the vagus nerve. We began calling this intersection of neuroscience, molecular biology, and biomedical engineering *bioelectronic medicine*, hoping one of the first successes from the new field would be new vagus nerve stimulating devices to treat inflammation.

Soon, I would meet the first person on earth who regained health through our first clinical trial.

A REVOLUTIONARY TREATMENT FOR RHEUMATOID ARTHRITIS

One sunny Sunday in November 2011, I flew to Sarajevo, the capital city of Bosnia and Herzegovina. A clinical trial sponsored by

SetPoint was in progress there to determine whether a vagus nerve stimulator could treat inflammation in patients. Nearing the airport on final approach, I pressed my forehead against the window and saw we were flying low over graveyards, field after field of countless white markers set against the greens of Bosnia's undulating hills, a reminder of the war this place had recently endured.

As we landed and taxied to the terminal, I redirected my thoughts to the country's now leading role in testing a new, experimental medical device. I imagined the vagus nerve stimulator rising from the devastation of war like a miniature phoenix destined to help alleviate suffering. Humans can do so much damage to one another but also so much good: I had traveled this far to meet the first patient in the first clinical trial testing this new idea of vagus nerve stimulation to treat inflammation. I wanted to see him with my own eyes, and hear his story in my own ears, because I believe what happened to him will rewrite the treatment of inflammation and the future of medicine.

Every major medical breakthrough begins with a single patient. Joseph Meister, at nine years old, had the good fortune of being the inaugural recipient of a rabies vaccine, a novel invention from the lab of Louis Pasteur. Before that happened, rabies had killed every human being it infected. It may be apocryphal, but some say Pasteur, reflecting on his legacy, requested a simple epitaph: "Joseph Meister lived." That Joseph Meister was the first patient and that he lived is not disputed.

Jonas Salk invented the polio vaccine and tested it on himself before giving it to strangers. He was the first, and his wife and sons were next. While further testing is always crucial, first patients are the breath of inaugural hope for revolutions in medicine.

The first patient to have a vagus nerve stimulator implanted to treat inflammation had been living with rheumatoid arthritis. Named from the Greek words for watery and inflamed joints, rheumatoid arthritis (RA) is a chronic autoimmune syndrome, meaning a condition in which the body's immune system attacks its own healthy tissues. Damage occurs when inflammation in the joints causes tissue destruction, leading to symptoms such as swelling, pain, and stiffness. The hands and wrists are often symmetrically affected, but RA's inflammation can also damage other body parts, including the skin, eyes, lungs, and heart. The exact cause of inflammation in RA remains elusive but is believed to stem from a mix of factors, both environmental and genetic. Think of it as a "friendly fire" incident. In this case, the body's white blood cells, called *monocytes*, *lymphocytes*, and *macrophages*, attack normal cells in the body. It's as if the immune system has come to the mistaken belief that healthy joints are infected with viruses or other germs and should be destroyed or eliminated.

Rheumatoid arthritis affects some twenty-five million people globally, including more than one million individuals in the United States alone. Women are more commonly affected than men, and the disease's onset is most frequent during middle age. Diagnosis is primarily symptom based, but X-rays and lab tests can offer further clarity. Beyond the joints, RA affects organs or other parts of the body in many cases, leading to complications like cardiovascular disease, osteoporosis, interstitial lung disease, and depression. But joint deformities, especially of the fingers, are particularly devastating. Timely and aggressive treatments can reduce the symptoms and decelerate disease progression, which helps people get by from day to day. But the therapies currently available are not cures.

Medications and pills (disease-modifying antirheumatic drugs, or DMARDs) target the inflammation, slowing its progression and reducing damage. Common examples of DMARDs include methotrexate, hydroxychloroquine, and sulfasalazine. Another class of anti-inflammatory drugs, called *biologics*, are derived from living organisms to specifically target the immune system. Biologics (also known as biologic drugs or biopharmaceuticals) are complex mixtures of proteins, sugars, or nucleic acids, or may be living cells and tissues, and must be administered using invasive methods, including needle injections.

Biologics, including monoclonal anti-TNF antibodies, work by targeting specific cytokines (small proteins that act as chemical messengers in the immune system) or other proteins and cells involved in the inflammatory process, thereby suppressing the immune system. These powerful immunosuppressive drugs can also cause dangerous side effects and come with a "black box warning," indicating that a potential side effect is death. They also have high costs and only benefit approximately 40 percent of rheumatoid arthritis patients.

We hoped to be able to offer a safer treatment for these patients that would be effective in a higher percentage of them. But first, we had to find out if vagus nerve stimulation would work in people with RA.

Ralph Zitnik, SetPoint's chief medical officer (CMO), and I settled into the back seat of a small European car for a bumpy two-hour drive. We were headed into the mountains to Mostar, a quaint town southwest of Sarajevo and, historically, the capital of Herzegovina. Mostar stretches along the banks of the Neretva River, combining historical importance and natural allure. The *Stari*

Most, or Mostar Bridge, with its majestic stone arch poised nearly eighty feet over the emerald river below, is more than just a marvel of sixteenth-century Ottoman engineering; it is a symbol of unity and a vital conduit on a centuries-old trade route. In fact, the town gets its name from the word for the original bridge keepers (*mostari*).

On our way to the hospital where the clinical trial was taking place, Ralph and I stopped at the bridge to watch high divers launch themselves off into the depths. We talked about its destruction during the war in 1993, then reconstruction and reopening eleven years later. It was both a restoration of stone and mortar and a symbolic rebirth arising from the indomitable spirit of Mostar. I was inspired and hopeful of soon witnessing another renaissance, one bridging centuries of vagus nerve research to a modern breakthrough for patients.

The University Clinical Hospital in Mostar (Sveučilišna klinička bolnica Mostar) sits on a high bluff overlooking the city about a mile from the river. Built in 1977 as a regional medical center, it suffered major damage during the war but, like the bridge, was rebuilt not long after. The stark, imposing rectangular building with high windows reflects a prioritization of function over form. Pulling into a wide circular driveway, we were met at the front door by Dr. Ante Bogut, one of the local physicians on the clinical team running SetPoint's trial there and the only one who spoke English.

Making no effort to contain my excitement, I thanked the doctor for his team's efforts in enrolling the first patient in history to receive a vagus nerve stimulator to treat rheumatoid arthritis and for giving me the chance to find out if my idea actually worked.

Dr. Bogut told me about the patient. As we strolled the hospital's long, beige-linoleum halls, I learned that Pero Dragoje was a father of two young children, a middle-aged man who had made a living as a truck driver. But the onset of rheumatoid arthritis had crippled him since his hands and feet hurt so much that he was unable to work. His doctors had treated him for years with prednisone, a steroid medication used to suppress inflammation, as well as with methotrexate, one of the DMARDs. That was all they had to offer because biologics like anti-TNF were not available in Bosnia then. So, having run out of treatment options and after spending years of long days in pain, lying on his couch, unable to work or play with his kids, he jumped (figuratively) at the chance to be in the SetPoint clinical trial.

Three months before I visited Pero and his clinical team, Dr. Richard Bucholz, a professor of neurosurgery at Saint Louis University, had traveled to Mostar to perform the surgery and implant a SetPoint vagus nerve stimulator on the first patient's left vagus nerve. Richard brought an early interest in computers to his training as a neurosurgeon, which, along with his natural dexterity, put him at the forefront of neurosurgery and digital technology. He helped develop the implant that he came to place in Pero Dragoje's neck, after his work with vagal nerve stimulation for patients with epilepsy and depression (see chapters 4 and 6, respectively). Richard estimates that he has implanted over two thousand vagus nerve stimulation devices so far.

At SetPoint, Yaakov had discovered that only a small amount of current was necessary to turn off inflammation during arthritis. So he modified the factory settings for devices used to treat epilepsy (which typically deliver five minutes of current alternating with five minutes of off-time around the clock, 24-7) to stimulate for only five minutes, just three or four times a day, and to deliver one-tenth of the current. Although these new settings blocked inflammation in the lab, until then, we did not know what would happen in patients.

Pero Dragoje was seated in a conference room with two physicians and a hospital electrophysiology technician. Rising when he saw me, his face a giant grin, he warmly, steadily, and firmly shook my hand, squeezing without evidence of pain or discomfort. He was big, about six foot four, with broad shoulders and a high forehead, and the grip of someone who had spent years hauling heavy crates and pallets. Although he did not speak English, with Dr. Bogut translating, I asked Pero about his battle with RA. He told me that at the time of the implant surgery when he was fifty-seven, his hands had become useless. He was unable to move his fingers or pick up things. And he feared things would never get better. Then he heard about the trial.

Since his vagus nerve stimulator was activated, he felt great, telling me that within a week of surgery, his hands and feet stopped hurting. And he was able to work again.

I said, "That's amazing!" Smiling, he said, "Yes!" and we all laughed.

I probed some more, inquiring whether Pero noticed anything else, like a change in his body weight or appetite or sleep habits, but he said no, all he noticed was that the pain stopped, and his hands, wrists, and feet all felt better. Within several weeks, completely pain-free, he had resumed driving his truck and handling heavy boxes of cargo. He went back to playing tennis, once so vigorously that he sprained his knee. This upset his physicians and caregivers, who gently scolded him and told him to take it easy, at least for the duration of the trial, since a swollen knee jeopardized his clinical score.

I was struck by the need for such advice to a man who just a few weeks earlier had been confined by pain and swelling to his couch, unable to do much of anything, least of all tennis.

We chatted for another half hour or so, then posed together for a picture. In the photograph, Pero towers over me, a full head taller (and it's a sizable head), but we are smiling. Many who have seen this picture point out that I am smiling more than he is, and perhaps I am. It was one of the happiest days of my professional life. Few medical inventors get the opportunity to meet the first patient who benefited from their idea. Afterward I spoke privately to the physicians, including Dr. Milenko Bevanda, the hospital's physician-in-chief and a highly respected professor who had mentored many of the leading Bosnian physicians in that era.

"So, it works, right?" I said.

Dr. Bevanda nodded. During the entire meeting he had spoken little, silently sizing me up as I conversed with the patient. I wondered if he didn't speak English, but I had noticed his scowl and his arms crossed over his chest in a defensive posture, so I guessed that the experiment's positive outcome had surprised him.

Undaunted, I said, "You thought this was a crazy idea that could never work, didn't you?"

He admitted in perfect English, "I thought there was no chance at all this would work." We all laughed together for several minutes as we shook hands and bade a warm goodbye.

Thirteen years later, as of this writing, Pero's device is still working. He is feeling great, "functioning without any limitations," I am told, and long-hauling it around Europe for a living.

On some days his hands feel stiff in the morning, but minimally. It occurs to me that Pero is seventy now and some stiffness could well be within the expected range for his years. Two years after the study was completed, Pero stopped taking any medication for his rheumatoid arthritis. Once in a while, after playing sports or doing some gardening, he might have an Advil or other nonsteroidal anti-inflammatory. As Pero says of his vagus nerve stimulator, "I would be a dead man without it."

After that first meeting with Pero, I tendered my resignation to the board of directors at SetPoint. I received several phone calls from board members questioning my decision. For me, the answer was simple. After working on this idea for more than fifteen years in the lab, then seeing the success of the first patient in the first clinical trial, I knew it was indeed possible to stimulate the vagus nerve in humans to prevent or reverse inflammation. When we started SetPoint, my fervent wish had been to establish an environment where the ideas from my lab could be tested in a clinical trial. Now this mission was accomplished, and I wanted to return to my laboratory and the Feinstein Institutes to do new projects, make new discoveries, and invent new cures.

Scientists who are fortunate enough to succeed at making discoveries that can be turned into treatments are not content to accept the world as it is. They want to understand how things work, push the boundaries of knowledge, and turn that knowledge into useful inventions that benefit humanity. This process can take many years, even decades. It takes persistence, dedication, and a willingness to fail. And it requires building new environments, which is why creative scientists often build new institutions. Scientific research is collaborative, and the best chances for making new discoveries happen by bringing the best and brightest together in a supportive and well-resourced setting. Jonas Salk developed the first effective polio vaccine, and he founded the Salk Institute for Biological Studies to continue his research and train future generations of scientists. Louis Pasteur developed vaccines for rabies and anthrax, and he founded the Pasteur Institute, now a global network of research centers. So, in November 2011, still flying high from my trip to Bosnia and Herzegovina, I returned to the Feinstein Institutes, to my lab and to my colleagues, and redoubled my efforts.

FROM A STUDY OF ONE TO A STUDY OF EIGHTEEN

In addition to the first patient in Bosnia, we also studied seventeen other patients with rheumatoid arthritis. All of them had active disease, meaning they experienced painful swollen joints and some level of impairment or disability despite receiving methotrexate, a DMARD, for a minimum of three months. Richard Bucholz and several other neurosurgeons performed vagus nerve stimulator implantation surgeries in Mostar, Bosnia; Zagreb, Croatia; and Amsterdam, the Netherlands. Two weeks after the surgery, the study investigators increased the electrical current parameters based on the individual patients' ability to tolerate it. As current increased, some subjects noticed a tingling in their voice when they spoke (the recurrent laryngeal nerve effect). Other subjects noted twitching or tingling of the muscles in their neck, but most of these side effects are short-lived and well tolerated.

Within several weeks, all the subjects tolerated the stimulation therapy, averaging 1.3–1.6 milliamperes of current delivered in five-minute bursts three or four times daily. The effect of vagus nerve stimulation on inhibiting TNF production was significant: After implanting the vagus nerve stimulator, we found that TNF production declined by nearly 50 percent. Even more important for the patients, the signs and symptoms of their rheumatoid arthritis significantly improved.

Five years later, in July 2016, we published an article in the *Proceedings of the National Academy of Sciences* detailing the results of SetPoint's clinical trial. Our headline, in the neutral voice of science, belied our excitement: "Vagus Nerve Stimulation Inhibits Cytokine Production and Attenuates Disease Severity in Rheumatoid Arthritis." But behind this curtain of our professional mien, there had been high fives and a champagne celebration. While my lab had discovered many years earlier that the vagus nerve controls inflammation by suppressing TNF and other cytokines in mice, this clinical study was the first time anyone had investigated whether vagus nerve stimulation could do this in humans.

Recall that inflammation is implicated in two-thirds of human deaths globally, and it is also a leading cause of illness and suffering among the living. Remember, the first patient was able to return to work and to play with his children again. To this day, we do not believe our high fives were premature.

As I was writing this chapter, a pivotal trial was just completed in the United States, where the FDA approved a so-called "breakthrough designation" for a new VNS device developed by SetPoint Medical. Called the *RESET-RA study*, this trial assessed the safety and efficacy of a new *SetPoint System*, a Tylenol-sized device surgically embedded onto the vagus nerve through a small incision on the neck's left side. The study earmarked spots for 250 participants across forty locations in the United States to assess whether it provides relief to patients with active, moderate to severe rheumatoid arthritis—particularly those unresponsive or intolerant to biologic or targeted synthetic DMARDs. Every participant, upon qualification, got the implant, but post-implantation only half experienced active stimulation, while the other half served as a control group. That is, those in the control group didn't know that they weren't receiving vagus nerve stimulation. After twelve weeks, the control subjects transitioned to active stimulation for a 180-week follow-up with both groups—treatment and control—receiving active stimulation.

Then, by the time I was working on manuscript revisions, the results of the RESET-RA study came in, and they were good news indeed. The SetPoint VNS system is the first neuroimmune modulation device to demonstrate a clinical benefit in adults living with moderate to severe rheumatoid arthritis. My hope is that these favorable results, providing essential evidence of the treatment's efficacy, safety profile, and long-term outcomes, will soon lead to FDA approval for a new therapy benefitting millions of future patients.

According to one study (and there are many others like it), 74 percent of rheumatoid arthritis patients are extremely unhappy with their treatment options. Pain, fatigue, and insomnia continue to cripple their quality of life. Study respondents experience "either moderate disease activity (37%) or high disease activity (33%). . . . Very few achieved remission (16%) or low disease activity (13%) with their current treatment. Of those individuals who reported not being satisfied with their treatment, approximately half were currently experiencing a flare (a period of exacerbated symptoms)." I hope more people suffering from rheumatoid arthritis will gain access to this new, science-backed treatment.

TREATING CROHN'S DISEASE

The first patient I met with Crohn's disease was Margie, a newlywed twenty-nine-year-old second grade teacher from West Hartford, Connecticut. Like many Crohn's patients, her symptoms began when she was in college. Sometime during her senior year, she started having bouts of severe abdominal pain and diarrhea. These episodes lasted weeks at a time and persisted after graduation, and by her thirty-first birthday, her medicine cabinet was overflowing with failed remedies. Despite countless medications and dietary modifications, nothing helped.

Things took a serious turn for the worse for Margie when she developed a high fever, chills, vomiting, and debilitating abdominal pain. At St. Francis Hospital in Hartford, Connecticut, she drank barium prior to having abdominal X-rays, a test used to assess the flow of liquids through the bowel. Her surgeon diagnosed strictures, or narrowing, meaning the normally soft, pliable intestinal tubes were pinched in places to the width of a pencil, too small for food and digestive juices to pass. At surgery, he removed the inflammation and scar tissue from her small intestines, afterward giving Margie her diagnosis.

Named for Dr. Burrill Crohn, a gastroenterologist in New York City, Crohn's is an inflammatory bowel disease, an autoimmune condition where inflammation affects segments of the gastrointestinal tract from the mouth to the anus, including the esophagus, stomach, duodenum, and small and large intestine.

Since Crohn's first description, in 1932, other signs and symptoms have been added to the list: skin rashes, arthritis of the extremities and spine, eye inflammation (*uveitis*), ulcerations of the skin (*pyoderma gangrenosum*), and inflammation of the fat underneath the skin (*erythema nodosum*) among them. As in Margie's case, complications of excessive inflammation may cause bowel obstruction, strictures, and scarring, but also abscesses and fistulas which drain bowel contents through the skin or into other body compartments.

After striking in a person's late teens or early twenties, Crohn's disease often begins an inexorable march through young adulthood, punctuating life with months of abdominal distress, diarrhea, fatigue, weakness, and incapacitation. Steroids (prednisone) and other anti-inflammatory drugs help some patients, but for others they provide only short-term periods of relief before intra-abdominal complications recur, requiring surgery in as many as 50 percent of patients. Making matters worse, steroids have serious side effects, including weakness and wasting of muscles, fatigue, cataracts, and fat retention. And because steroids suppress the immune system (which is how they suppress inflammation), this also renders patients susceptible to other potentially life-threatening infections. Immunosuppression means their immune systems cannot respond normally to defend them against dangerous viruses and bacteria.

I was nine years old the first time I stepped foot in a hospital, and the occasion was to visit Margie, my new stepmother. After my mother died, my father moved us to West Hartford, Connecticut, to be closer to our large extended family. There he met Margie on a blind date, fell in love, and they married. Attempting to reassure me from her hospital bed, Margie said she felt fine and would be home soon. But I was filled with questions. What caused this disease? How could it be treated? When would the next attack occur? For the second time in my life, I was frustrated by my inability to help my sick mother. And by not understanding the cause of her illness.

When she was not in the hospital, Margie encouraged me to take things apart and rebuild them, but whether this was a ploy to keep me out of her hair or to guide my development as a future physician-scientist, I cannot know. As a youngster, I began with battery-operated toys that I took apart to see what made them work, a practice that later progressed to motorized toy cars and airplanes and then to lawn mowers, radios, washing machines, and eventually real cars. Without telling my father, who would have said no, Margie lent me the money to buy my first car. By this time, my siblings and I called Margie *Mom*, and to me she was the greatest woman in the world.

Although it has been more than fifty years since I first saw the inside of a hospital while visiting Margie, we still do not know what causes Crohn's disease. However, the most recent results of SetPoint's clinical trials indicate that vagus nerve stimulation may be a useful therapy. This clinical study, which I coauthored, assessed the therapeutic effects of vagus nerve stimulation in sixteen patients, including Kelly Owens from chapter 1, with moderately to severely active Crohn's disease, using study endpoints based on the Crohn's disease activity index (CDAI). This standard assessment tool is based on a combination of patient-reported symptoms and clinical measures.

As in SetPoint's rheumatoid arthritis studies, patients received vagus nerve stimulation for only twenty minutes or less daily, but within sixteen weeks there was a significant decrease in CDAI as well as objective evidence of a decrease in inflammation as observed in the intestinal lining by endoscopy in eleven out of fifteen of the patients. Once again, the TNF and other cytokines associated with inflammation were also significantly decreased. In this trial, too, vagus nerve stimulation was not only safe and well tolerated but resulted in a meaningful reduction of clinical disease severity and an improved quality of life. Patients became healthier and happier.

Today we understand that a specific subset of vagus nerve fibers has a critical role in turning off or turning down the inflammatory response in mice and people, restoring balance (homeostasis) to the immune system. Low levels of electrical current applied for only a few minutes, a few times a day, stimulate this anti-inflammatory reflex. And it produces a lasting effect that persists for twenty-four to forty-eight hours, meaning that less than twenty minutes of electrical stimulation per day should be sufficient to control inflammation in most patients.

Even though vagus nerve stimulation was not available for Crohn's disease when Margie needed it, she continued to cheer me on until her dying days and say her prayers that this therapy might one day offer relief to others. I expect that vagus nerve therapy, like every other medical therapy, will not work 100 percent of the time in 100 percent of patients. But after decades of using vagus nerve stimulation to treat other conditions, including epilepsy, for which hundreds of thousands of patients have already been treated with an implanted device, it is a safe bet that this therapy is relatively safe. It is certainly safer than many of the other invasive treatments for rheumatoid arthritis and inflammatory bowel disease, like anti-TNF biologics, which I had a hand in helping launch into current use. Again, anti-TNF and other biologic agents carry black box warnings and are very expensive: They can cost upward of $50,000–$100,000 per year for life. By all indications, a lifetime of vagus nerve therapy, which does not have a black box warning, will cost about the same as only one year of biologics, and a single minor surgery will supplant the need for years of costly, injectable, potentially toxic biological therapies.

I keep a vagus nerve stimulator, which is the size of a Tylenol capsule, on my desk to show people who might not otherwise believe it, and to reflect, in quiet moments, on the immense promise of such a tiny thing. I've been working on inflammation for decades—it's been more than twenty years since I registered the invention with the United States Patent Office—but this thing still amazes me.

After decades of clinical testing—and improving many people's lives—as well as advances in the science and technology, vagus nerve stimulators may finally become broadly available for treating the inflammatory threats looming over humanity. How many of these diseases might vagus nerve stimulation neutralize? Can it add years, or perhaps decades, to healthful life? We are starting to find out, because a growing body of evidence continues to indicate that vagus nerve reflexes can promote recovery by suppressing inflammation.

This new field of bioelectronic medicine is only just beginning, but hopeful glimpses of a better future shine through in the experiences of patients from completed and ongoing clinical trials. People who were once sidelined or even crippled by inflammation have rediscovered vitality. And we are now at a tipping point when VNS may be able to help you or your loved ones. If you are suffering from diseases triggered by inflammation, I hope this chapter has shown you that vagus nerve stimulation therapy offers real hope. See chapter 11 for guidance on how to talk to your doctor about whether VNS is right for you.

6

Beyond Medication

A Healing Reflex for Depression

No amount of love can cure madness or unblacken one's dark moods. Love can help, it can make the pain more tolerable, but, always, one is beholden to medication that may or may not always work and may or may not be bearable.

— KAY REDFIELD JAMISON

A lot of hatred, that's how I felt. I hated everybody and everyone," Nick Fournie remembered. "You know, it was always being in a bad mood, down, nothing good. Even if I had everything that I wanted, it wouldn't make me happy."

Nick was depressed, sleeping during the daytime as well as at night to escape his suffering. "It was like every day was two days," he said, and sleep was the only way he could bear for time to pass at that crawling, maddening pace. Antidepressants gave him no relief and, worse, sometimes produced the opposite, rendering him even more tired and disinterested in life. It was like this for years until Nick received his vagus nerve stimulator.

A very few of us are always happy, every day finding joy in the bustling, beautiful world and reveling with gratitude for life's abundant gifts. Most of us are content enough, enough of the time, to say we are basically okay, doing our best and muddling through a mix of blessings, disappointments, achievements, boredom, joys, and the occasional horror, perhaps with the help of psychotherapy or psychopharmaceuticals or both. But at the other end of the spectrum, some of us find ourselves in a dark tunnel of misery with no end that we can see, mired in a paralyzing gloom that reduces life to a tiresome burden with no respite in activities and friendships that used to lift our spirits. And sometimes, as in Nick's case, psychotherapy and drugs don't help at all or enough.

For reasons that we do not well understand, depression is common. It is becoming more so, with the World Health Organization reporting that up to 280 million people suffer from serious depressive disorders. But that estimate likely undercounts the true number of depressed individuals since many patients avoid admitting their symptoms to their healthcare provider, making diagnosis difficult and lowering reported incidences. What I do know, even without knowing you personally, is that someone you are close to, a family member, friend, or colleague, or perhaps you yourself, is depressed or has in the past experienced depression. That's how common it is.

As one of nearly six hundred participants in a clinical trial at Washington University School of Medicine in St. Louis, Missouri, Nick Fournie received a small, circular device not much bigger than an American silver dollar implanted under his left collarbone. It sends tiny blasts of electrical current through a wire connected to his left cervical vagus nerve, stimulating the nerve on a recurring cycle that repeats every five minutes: Thirty seconds on, five minutes off, the device operates twenty-four hours a day. It's like a pacemaker for the brain.

The participants in this study, like Nick, had depression that could not be alleviated by four or more antidepressants, taken either separately or in combination. As many as two-thirds of the fourteen million Americans with clinical depression aren't helped by the first antidepressant drug they are prescribed, and up to one-third don't respond to subsequent attempts with other such drugs. Dr. Conway's team followed 328 patients implanted with vagus nerve stimulators, many of whom also took medication. They were compared with 271 similarly resistant depressed patients receiving only treatment as usual, which could include antidepressant drugs, psychotherapy, transcranial magnetic stimulation, electroconvulsive therapy, or some combination.

The study was designed to evaluate the quality of life of patients who were just barely getting by despite taking multiple antidepressant medications. It found that when they added vagus nerve stimulation therapy, it made a significant difference in the participants' everyday lives. Among the fourteen categories of quality of life that the researchers evaluated were physical health, family relationships, ability to work, and overall well-being. And on ten of the fourteen measures, those with vagus nerve stimulators did better. For a person to be considered to have responded to a depression therapy, they had to experience a 50 percent decline in their standard depression score. However, the researchers noticed that some patients with stimulators reported they were feeling much better even though their scores were only dropping 34 to 40 percent.

Nick says he only notices the sensation of the device being "on" for those thirty seconds every five minutes if he is "real calm and real still." When he's busy, he doesn't know it's there.

Embracing his waking hours after getting his VNS device, Nick and his wife, best friend, and high school sweetheart, Mary, who supported him through the trenches of depression and encouraged him to get the implant, say his severe depression is behind them. "We enjoy every day together and I enjoy life." Cue a YouTube video of Nick driving a motorcycle, Mary on the back, cruising into the proverbial sunset, a vagus nerve stimulator invisible under his skin on their journey back to homeostasis.

The vagus nerve stimulator that Nick and the other study participants received is like the devices used to treat epilepsy. My friend Bill Bell, the neurosurgeon you met in chapter 4 who placed the first implantable VNS device in Toney Kincaid, has also done these procedures for depression patients. He recalls a young woman in her twenties, whose family approached him in desperation. They had tried everything, including maintenance electroconvulsive therapy, and had "been everywhere, the Mayo Clinic, you name it," Bill said. Nothing gave her much relief until vagus nerve stimulation with her new implant "was like turning a switch." Bill says he got Christmas cards from the family for years after that. One case doesn't prove that it works, Bill is careful to note, but "it sure was wonderful."

Of course, Bill is right. One case doesn't prove efficacy. But as a physician-scientist, I can add that it is equally true that first patients, like Joseph Meister (the first person to be vaccinated against rabies) or Toney Kincaid and Pero Dragoje (the first two people to receive implanted vagus nerve stimulators for, respectively, epilepsy and RA), encourage us to study more patients.

Hundreds of thousands of people have implanted vagus nerve stimulators for epilepsy, and interestingly, even when it didn't help with their seizures, they opted to keep their implant because it improves their mood.

However, despite the many other people whose stories of depression are similar to Nick's and Bill's patients, and despite FDA approval of implanted vagus nerve stimulation devices for depression since 2005, the therapy is not widely used. This is due to several reasons, including variable insurance reimbursement policies and a relative lack of clinical data. Another significant barrier to wider adoption is that we do not know why it helps some but not all patients and because, at the most fundamental level, we do not understand what causes depression.

There are many theories of depression, too many to cover in this discussion, but let's consider some that might point to a role for the homeostasis-maintaining functions of the vagus nerve. And let's consider how vagus nerve reflexes could be involved in healing depression.

BRAIN OR BODY, OR BOTH?

The brain is the central focus of depression research and therapy for the obvious reason that it is the primary organ responsible for processing sensations, emotions, and thoughts. Mental health professionals recognize depression as a complex syndrome with no single cause, featuring a wide range of symptoms from persistent sadness and loss of interest in daily activities to sleep and appetite disturbances, fatigue, and impaired concentration. A diagnosis of "major depressive disorder" adds to this list feelings of helplessness, despair, and paralyzing exhaustion. Its hallmark, psychomotor retardation, manifests as an overwhelmed mind and body, like Nick struggling with simple tasks of daily living.

Modern neuroscience has revealed the brain to be a tangled neural circuitry buzzing with one hundred billion individual neurons. These neurons fire specific action potentials that communicate via trillions of synapses, underlying our very thoughts, emotions, and sensations. Brain imaging studies can illuminate the electrical and chemical activities of specific brain centers in depressed and nondepressed individuals. While these studies show some differences between depressed and nondepressed individuals, it is difficult to determine how these brain activities relate to personal experiences, raising as many questions as answers about how specific neural circuit malfunctions cause depression.

Like trying to pick out a starry constellation on a mostly cloudy night, we look at the brain with powerful new tools but only seem to catch glimpses of insight in our efforts to explain depression. The cause—or more likely, causes—are hidden deeply within a vaster unknown. Could evolution, we wonder, which adapted and molded the human brain for optimal survival and success, have planted the seeds of despair within?

It is indeed possible that our modern tendency to develop depression has roots in a prehistoric time when early humans began to use language. The ability to communicate using words bolstered social bonds and cooperation, enhanced reproductive success, and fostered the expansion of tight-knit communities. But this was also a setup for painful disappointments, rejections, and isolation because the very language that aided our ancestors' survival could also produce hurt feelings, dashed hopes, and social distress.

Closely linked to early language was another mixed blessing:

a newfound ability for future planning. Forward planning enabled early humans to create more sophisticated tools, develop long-term hunting strategies, and establish seasonal camps. But once again, this came with a trade-off. Planning for the future, while beneficial, may have paradoxically contributed to anxiety and depression. Planning can escalate into excessive worry when future events are perceived as threatening, a kind of rumination in which the mind exists in a state of heightened alertness and anxiety focusing on negative outcomes rather than present realities.

Millions of years later, with talking and planning capabilities that our predecessors until relatively recently couldn't imagine, we go about our daily routines never worrying about charging lions, and only rarely about angry dogs. We have come a long way from our evolutionary ancestors who spent much more time thinking about such things, along with the associated escape routes of, literally, fighting or flighting. But there is bad news, too, because modern technology has evolved new methods to torment us with a relentless influx of notifications and alerts from ever-present mobile phones, watches, and other devices to constantly remind us of our modern stressors and prod our sympathetic nervous systems. This constant barrage of information and alerts erodes our ability to recuperate, interferes with balanced sleep cycles, and lays the groundwork for heightened and sustained levels of anxiety, all of which contribute to depression.

However, when we recall that your brain and body are linked through nerves that communicate constant streams of information between them to maintain physiological balance in your organs and vital systems, we can also imagine that depression could arise from the body as well. As we have seen, the vagus nerve is crucial in maintaining a balanced state of mind and body through its reflexes.

But if your brain is stressed, anxious, or depressed, and your brain is intimately linked to your physiological health through the vagus and sympathetic nerves, then what does a stressed, anxious, and/or depressed brain state do to your bodily health?

A lot, and it's not good.

Recurring stress and anxiety increase sympathetic nervous activity. This floods your body with catecholamines, the neurotransmitters epinephrine and norepinephrine, which raise your blood pressure, heart rate, cardiac output, and breathing. You feel jittery and tense, ready to fight or flee.

These responses suppress your heart rate variability (HRV). As we learned in part 1, a low HRV means your sympathetic nervous system is in overdrive, exceeding the activity of your vagus nerve and parasympathetic nervous system. In other words, you are less able to rest and digest. HRV tends to be 20–30 percent lower in depressed individuals compared to nondepressed, and systolic blood pressure is 5–10 mm Hg higher in those with depression.

Stress responses also activate your adrenal glands to release *glucocorticoids*, hormones that stimulate *gluconeogenesis*, the production of glucose in the liver. This, in turn, increases your blood glucose levels. Elevated glucocorticoid levels, as occurs in depressed patients, accelerate *lipolysis*, the breakdown of fats into fatty acids, while suppressing digestion, muscle growth, and reproduction. Glucocorticoids also inhibit the action of insulin, meaning that your cells are less responsive to insulin. This further increases your blood glucose, sometimes even to dangerous levels.

These hormonal and metabolic responses are useful in a

short-term fight-or-flight situation, like escaping a lion or dog in a minute or two. But chronic activation of these responses is unbalanced and unhealthy. And prolonged exposure to high levels of glucocorticoids contributes to globally rising rates of type 2 diabetes, obesity, hypertension, and other conditions.

If that wasn't enough, chronic sympathetic activation also increases inflammation. The catecholamines released by the sympathetic nervous system bind to receptors on the immune system's white blood cells, stimulating these cells to produce more cytokines that increase inflammation. The catecholamines also sensitize white blood cells, priming them to respond aggressively when next encountering another stimulus. This state, called *trained immunity*, means that even a minor injury or infection leads to an overreaction, causing much more inflammation, and much more damage, than a normal response. When this occurs, your white blood cells will churn out powerful inflammation-producing cytokines, including TNF, IL-1, and IL-6, the same ones that are normally inhibited by the vagus nerve. In short, chronically increased sympathetic nervous system activity has the opposite effect of the vagus nerve. Instead of reducing inflammation and cytokines, it increases inflammation and disrupts your homeostasis.

According to numerous clinical observations, chronic stress increases inflammation and the risks of heart disease, diabetes, autoimmune disorders, and even cancer. Controlled experiments with human volunteers involving stress-inducing tasks show increased cytokine levels and other inflammatory markers in the blood and saliva following stress exposure. Since psychological stressors elevate cytokine levels, the brain's processing of thoughts, sensations, and emotions can manifest as inflammation in the body.

Just as stress in the brain can trigger inflammation in the body, inflammation in the body can also affect the brain since this connection between the brain and body is bidirectional. Once again, this has a significant impact, and it's not good. But this link between cytokines, inflammation, and depression could also be the key to unlocking why vagus nerve stimulation may help millions of people.

THE VAGUS NERVE AND SICKNESS BEHAVIOR

Which comes first, chicken or egg, inflammation or depression, is still debated. However, recent evidence from depressed individuals shows chronic inflammation appearing in their blood as elevated levels of inflammatory markers, like C-reactive protein (CRP) and inflammatory cytokines. Although many studies support this conclusion, consider that a review of published studies of 5,166 patients and 5,083 controls found significantly elevated levels of TNF, IL-1, IL-6, CRP, and many other inflammatory mediators, leading to the researcher unequivocally concluding that "depression is confirmed as a pro-inflammatory state."

To understand what this means, recall the last time you were inflamed. It may have been because you had the flu, COVID-19, a sprained joint or broken bone, or even just a bad cold. Can you remember how you felt when you had elevated levels of inflammatory mediators in your bloodstream?

You were likely tired, socially withdrawn, grumpy, uncomfortable, emotionally *labile* (unstable), and uninterested in and frustrated by daily activities. Depending on how much inflammation you were carrying around inside your body and blood, you were weak and lacked appetite, even for your favorite foods. All you wanted to do was sleep—just like Nick.

Simply put, your inflammation gave you a (hopefully short-lived) state of depression. Which came first? Sometimes, it's inflammation. I know this is true because in forty years of studying inflammation by administering pro-inflammatory cytokines to mice, rats, rabbits, dogs, pigs, baboons, and humans, I have seen what happens. They all get inflammation and a depression syndrome.

Cytokines produce signs and symptoms of depression in the brain because they activate sensory (input) signals in the vagus nerve, and because they can also enter the brain via the bloodstream after traversing openings in the blood-brain barrier. In animals, these incoming vagus nerve and blood-borne cytokine signals produce a constellation of signs called *sickness syndrome*, characterized by behavioral withdrawal, anorexia, weakness, lethargy, and other features that resemble depression in humans. Although we don't know much about the emotional experience of animals, because they cannot tell us how they feel, we do know what it feels like for humans who have been injected with cytokines. They develop depression.

As a young doctor training to be a neurosurgeon, I collaborated with Dr. Stephen Lowry to do clinical experiments by infusing a highly purified, FDA-approved bacterial endotoxin into healthy human volunteers. Most of these volunteers were marines and sailors on shore leave for two weeks in New York City, and they were quite happy to spend a day in the hospital with us and our infusions in return for a hundred-dollar stipend. By a protocol that was approved by the human ethical oversight committee at the New York Hospital, the endotoxin infusion we gave them stimulated their immune systems to release large amounts of cytokines, a syndrome that in today's post-COVID era is commonly known as *cytokine storm*.

What happened next was unforgettable. Within a few minutes after the cytokines appeared in their bloodstreams, they began to complain of "not feeling so good," with dizziness, fatigue, muscle aches and pains, lethargy, and loss of appetite. Then they became irritable and depressed as their increasing cytokine levels caused the onset of emotional withdrawal from activities that used to bring them pleasure. Later, in the lab, when we measured the serum levels of cytokines in their blood, we confirmed high levels of TNF, IL-1, and IL-6, the same cytokines that the vagus nerve normally inhibits.

Today there are still a few research centers studying the physiological and psychological effects of cytokines by administering endotoxin to human volunteers, but a more common circumstance for giving cytokines to humans is in treating inflammatory conditions and some cancers with biological drugs called *interferons* and *interleukins*. Many of these patients, prior to receiving cytokine infusions, are referred to a psychiatrist and treated with antidepressant medications in advance of receiving the infusions because we *know* that the cytokine infusions will cause some of these patients to become severely and clinically depressed.

In our research with the sailors, and with cancer patients, it is quite clear that the cytokines and inflammatory mediators are producing depression, not the other way around. The brain adopts a depressed emotional state in reaction to the appearance of cytokines.

So, when we experience chronic stress and anxiety, we may become victims of the depression that can result from the body's inflammatory response to psychological and emotional stress. And if we develop a chronic inflammatory condition, we may also risk developing depression. Clinical studies support this bidirectional (brain-body and body-brain) theory of depression because higher rates of depression are observed in patients with underlying inflammation caused by autoimmune diseases, cancer, prolonged immobility, or chronic infection as compared to the general population.

The back-and-forth causality of inflammation and depression is a bit of an intellectual whirlwind, but it makes more sense when we recall that the nervous system is much more than your brain dictating from on high. Indeed, your nervous system is a massive interacting network entwining body and mind. Neurons in your brain and neurons in your body are all connected. And the resultant behaviors they control are emergent, meaning they cannot be predicted by looking at one component, such as one nerve or one cytokine. The input signals arise first from every part of your body, not your brain. But they all travel to your brain, which processes its outputs to adjust your physiology and homeostasis accordingly. And your vagus nerve links all these networks together.

Although excessive inflammation is sufficient to cause depression, and inhibiting inflammation can reverse depression and improve quality of life for patients like Nick, let me say again: There is no one cause of depression in all patients. Treating inflammation with vagus nerve stimulation doesn't work for everyone, but because vagus nerve stimulation can suppress inflammation, I believe it will benefit some whose excessive inflammation in their body drives the depressed state in their brain.

This reasoning extends to other common environmental risk factors for depression, including psychosocial stress, early life adversity, obesity, and a processed-food diet—all of which are pro-inflammatory—since exposure to pro-inflammatory cytokines induces a sickness syndrome with symptoms that overlap with those of depression. Now, how does this inflammatory theory of depression and vagus nerve reflexes line up with the leading pharmacological theories of depression? It's a very good question that I wish more researchers were asking.

For more than fifty years, a predominant theory of depression holds that it is caused by a "chemical imbalance" in the brain, blamed on low serotonin levels or impaired activity of the brain's serotonin receptors. It's such a popular theory that you've probably heard of serotonin even if you've never been diagnosed with depression. In this theory, *serotonin*, a neurotransmitter in the brain, is supposed to play a key role in regulating mood, digestion, sleep, and other bodily functions. Selective serotonin reuptake inhibitors (SSRIs) are drugs designed to increase the levels of serotonin in the brain, and their effectiveness at treating depression varies among individuals. On average, about half (40–60 percent) of patients taking SSRIs experience a significant improvement in their depression symptoms. The benefit depends upon the severity of depression, individual differences in biochemistry, and the presence of co-occurring mental health conditions. Fueled by this clinical evidence, pharmaceutical industry marketing, and the endorsement of major academic and government institutions, the serotonin theory gained near universal acceptance and led to skyrocketing use of antidepressants in recent decades. In the United States, antidepressants developed and marketed on this basis rank among the top three most common classes of prescribed drugs.

Nonetheless, during this same period of more people using SSRIs, the incidence of depression has increased, not decreased. And new studies are challenging the dogma of serotonin chemical imbalance as the cause of depression.

If an SSRI has helped you or someone you know, that is wonderful. Large randomized clinical trials of SSRIs indicate they confer some clinical benefit in some patients. But these results, and your personal experience, do not prove causality or confirm that serotonin dysfunction is causing depression. For example, SSRIs may also inhibit inflammation. The questioning and the answers that interest me most are those that could lead to new therapies that help more people, not fewer. Let's explore this idea a bit further.

A comprehensive review of the world's medical literature examined key research areas related to serotonin in depression. The review included seventeen studies covering thousands of patients, but there was no association between levels of the serotonin metabolite 5-HIAA and depression, or between levels of plasma serotonin in participants with depression. It concluded that reduced serotonin might be a consequence of antidepressant use rather than a cause of depression. Regarding serotonin receptors and transporters, evidence was weak and inconsistent, and the impact of prior antidepressant use on the measure of these effects was also unclear.

Genetic studies on the SERT gene, which encodes a transporter that moves serotonin out of the synapse, found no association between the SERT gene and depression or stress. The authors conclude that there is no consistent evidence supporting the hypothesis that depression is caused by lowered serotonin activity or concentrations, and on the contrary, some findings suggest that long-term antidepressant use might reduce serotonin levels. This is a direct challenge to the traditional serotonin hypothesis of depression.

Again, these findings do not mean that the 40–60 percent of patients who do feel better after taking SSRIs aren't gaining some real benefit. I am all for patients' benefits, especially with minimal side effects, and I spend my life studying how to implement new therapies for that reason. It's just that no one actually knows how these drugs sometimes make people with depression feel better, and sometimes make them feel worse. Interestingly, administering SSRIs to animals and patients with inflammation after receiving cytokines in the lab can alleviate depression caused by these cytokines. This anti-inflammatory role of SSRIs is little studied and incompletely understood, and I sincerely hope that my colleagues are inspired to investigate it further.

But whatever SSRIs are doing, it seems they can impart some protection against the depression-causing side effects of cytokines and other inflammatory mediators. This is why SSRIs are often used in clinical studies of cancer patients receiving cytokine treatment, when the risk of developing cytokine-induced depression is high. As in your own experiences with the flu or other times when you just haven't felt like yourself, the onset of depression can be confused with the development of sickness, because the symptoms associated with infections (or the administration of cytokines) mimic those of depression.

A HEALING REFLEX TO FEEL BETTER

If your vagus nerve suppresses inflammation, and inflammation causes depression, naturally, you might ask whether the vagus nerve can also be involved in causing the onset of sickness symptoms—or prevent them?

The answer is yes.

We know this from a series of remarkable experiments by Linda Watkins in the early 1990s that shed light on a surprising connection: the link between the vagus nerve and our body's sickness response. Intrigued by how infections and inflammation triggered behavioral changes in rats, Watkins sought to understand the mechanism behind fever, loss of appetite, and social withdrawal, where rats retreated to cage corners, becoming almost unresponsive after receiving an injection of an inflammatory agent (the cytokine IL-1).

Wondering if these signals reached the brain via the bloodstream or the vagus nerve, she designed an ingenious experiment. A group of rats received the same injections but with their vagus nerves severed. The result was striking. These animals lacked the typical fever and sickness behaviors, pointing to a clear conclusion: Injected IL-1, a key inflammatory molecule normally produced during infection or injury, triggered sensory nerves in the vagus nerve, relaying crucial information about the body's internal state to the brain stem. It was this communication, translated by the brain stem, that produced the cascade of responses, including fever, lethargy, loss of appetite, and withdrawal—or didn't when the communication couldn't happen because the vagus nerve was disabled. In essence, the vagus nerve, acting as a relay, incites the brain to orchestrate sickness behaviors in response to IL-1's presence. Watkins's discovery not only unveiled the vagus nerve's pivotal role in illness responses but also opened doors for novel therapeutic approaches targeting this crucial vagus nerve communication pathway.

Recall, the vagus nerve is not a solid wire; it is one hundred thousand individual wires or fibers (on each side), with 80 percent of these being sensory, that is, the source of input to the brain from the body. If the presence of an inflammatory agent like IL-1 hijacks a few hundred fibers to induce depression-like symptoms, that leaves tens of thousands of other fibers that could be involved in other things going right, or wrong.

As we saw in chapter 3 in our discussion of eating imaginary pizza, the vagus nerve supplies crucial reflex controls to the gut, which is a teeming metropolis of your bacterial microbiome comprised of more organisms living there than the number of cells you have in your entire body. Chronic stress, which, again, afflicts many living with depression, weakens the protective intestinal barrier that normally prohibits your gut microbes from activating your own immune system. When this barrier is breached, a condition known as *leaky gut syndrome*, bacterial toxins in the gut like lipopolysaccharide (LPS) slip through the cracks and infiltrate your gut wall and bloodstream, turning on the release of cytokines, including IL-1 and TNF. Lipopolysaccharide leaking from the gut can encounter those white blood cells that we just saw amped up by chronically elevated catecholamine levels, and the result is, as you can guess, even more inflammation.

All these inflammatory molecules—LPS, IL-1, and TNF—stimulate sensory vagus nerve fibers. These fibers travel to the brain and induce depression-like sickness responses. So begins a stream of incoming gut-to-brain vagus nerve signals that reach the brain stem before they are relayed farther up to the brain's emotional control center, the amygdala. A vicious gut-vagus-brain cycle of depression ensues, with the stress and pain of an inflamed gut to deepen the emotional dysregulation and perpetuate the cycle. Although still in the early stages of clinical studies, there is ongoing research addressing questions linking gastrointestinal inflammation to depression through probiotics and vitamins A and D. Other studies are exploring whether these and other modalities that may quell the inflammation-related signals from gut to brain can alleviate depression.

Perhaps electrical stimulation of the correct vagus nerve fibers, when we identify them, would have a similar effect of regulating inflammation in the gut and, in turn, alleviating depression. Several labs, including mine and Sangeeta's, are actively studying this idea.

While inflammation may be a cause or contributor to depression in some (or many) patients, there are many other reasons to study whether the vagus nerve contributes to depression via other mechanisms. I'll leave the details in the endnotes and make the larger point here. It will not surprise you, if you read part 1 of this book and appreciate the vagus nerve's reach into the brain and into all the body's organs and vital systems, that the vagus nerve, by carrying neurons that link countless functioning neural networks in the body and brain, could have many different roles in a complex condition like depression. In other words, if we consider depression as a failure of homeostasis, it is plausible that dysfunction of specific vagus nerve reflexes may contribute to the physiological and psychological imbalances of depression. And on the other side of the coin, perhaps vagus nerve stimulation can coax those reflexes into restoring mental and physical health and well-being.

The genesis of vagus nerve stimulation as a treatment for depression is a tale of serendipity and scientific inquiry, and we have come a long way from the observation that vagus nerve stimulation made some epilepsy patients feel better to the point where there are devices available for patients with severe depression. Ongoing clinical trials continue to explore the efficacy of vagus nerve stimulation in treating this all too common condition.

Early clinical trials have focused on patients with treatment-resistant depression, those who had not responded to conventional therapies, and a notable percentage of participants showed a significant reduction in depression symptoms. For instance, one pivotal trial reported that approximately one-third of patients experienced at least a 50 percent reduction in the severity of their depression. These findings, underscored by the fact that improvements with vagus nerve stimulation therapy were sustained over long periods, led to the FDA approving vagus nerve stimulation as a therapy for treatment-resistant depression in 2005. Although this marked a significant milestone in psychiatric treatment, VNS is far less common as a treatment for depression than for epilepsy.

This may change someday. Currently, a much larger clinical trial is underway, aiming to enroll up to one thousand patients who will be monitored for five years. During this time, one half of the subjects will receive active vagus nerve stimulation during the first year, and the other half will receive vagus nerve stimulation after a delay of one year. The pending results, if positive, could lead to broader reconsideration for the use and insurance reimbursement of vagus nerve stimulation therapy in treating severe depression.

Meanwhile, since that first FDA approval, patients like Nick Fournie have enjoyed remarkable results from vagus nerve stimulation for depression, rebuilding their lives, regaining their ability to work, and restoring their personal relationships. In some cases, this therapy has even prevented suicide, offering a lifeline to those on the brink of despair. But vagus nerve stimulation is not a magic bullet for depression or anything else. It is not effective for everyone, and its success largely depends on the individual's specific condition and response to standard therapies. Ongoing research into the role of inflammation and other mechanisms of depression should help us better understand the effectiveness of vagus nerve stimulation and identify those patients most likely to benefit from it.

As Nick put it, the effect can be "awesome." To me, that means worth further study, and worth making available to more people.

7

Outside-In Stimulation to Regulate Body Weight, Treat Diabetes, and More

Everything in excess is opposed to nature.

— HIPPOCRATES

On occasion, even though my own body mass index (BMI) is in the normal range, I have thought about losing a few pounds. Who hasn't, especially as we find ourselves surrounded by cultural messages that idealize fitness and thinness in a land of food excess? But thinking about it is not the same as doing it. So, although I hate to disagree with Ralph Waldo Emerson, "what you think you become" does not seem to apply to losing weight. In the end, I rely on scientifically proven guidance for weight loss (or weight regulation): getting a balanced diet, regular exercise, and enough sleep.

Yet, as we are learning more about how the vagus nerve maintains homeostasis, the nagging question remains: *Why* is it so difficult? Why can't we all just lose weight when we want to? When we need to? Maybe we can.

Because we knew that the vagus nerve reaches into the gastrointestinal and endocrine systems, regulating the reflexes of hunger and digestion, it seemed an ideal time for our lab to dive into the science and physiology of body weight. It was 2018, and almost immediately, we discovered two things. First, the vagus nerve reflexes that balance your weight, defending your body against weight loss and weight gain, are exceedingly complex. And second, no one, anywhere, completely understands how they all work. So, armed with questions but no good answers, we started to plan a few simple experiments and fatten up some mice.

During one weekly lab meeting, I took an informal survey to learn that 100 percent of my colleagues were also strongly in favor of losing a few pounds. But they wanted it to be painless and healthful, and they did not want to have to exercise for two hours a day. Nor did they want to fast. (I told you they are a smart crew.) I was, of course, pleased that they would rather hang out in the lab instead of the gym.

Then we talked for almost two hours about the many ways the vagus nerve might be involved in the homeostatic mechanisms of appetite, digestion, and body weight. We discussed how it transmits signals to the brain that cause satiety, regulate glucose and lipid levels, and decrease the inflammation underlying insulin resistance and type 2 diabetes. At the end of the meeting, we cleared away the empty pizza boxes (the irony was not lost on me) and returned to the lab. Then, Sangeeta designed and launched the new experiments to stimulate the vagus nerve in our obese mice using a noninvasive device, hoping to cause them to lose a few grams—the happy murine equivalent to a few pounds in a human. What happened next may point the way to a noninvasive, drug-free, do-it-yourself solution for everybody wishing to lose weight in the future.

Surgically implanted vagus nerve stimulators have significant advantages for many patients, especially as they ensure compliance with the planned therapy. Most people don't realize how elusive full compliance is when a drug or other therapy is put in patients' hands. We get busy, we forget, and we don't always do what's best for us. (Recall, honestly, the last time you were prescribed a course of antibiotics: Did you really adhere to the schedule and take every single pill on time for fourteen to twenty-one days?) For some conditions, patients and physicians are likely to prefer an implanted device as a reliable way to deliver therapy because it never forgets.

But for other conditions, such as elective weight loss, there may be a preference for devices that work through the skin, without the need for surgery. Bioelectronic innovators have been looking for ways to access the vagus nerve without surgery, and they have found some. This chapter explores possibilities for noninvasively stimulating the vagus nerve for obesity and weight loss, diabetes, and other conditions, focusing on those methods that are supported by solid scientific evidence.

TREATING OBESITY WITH ULTRASOUND

According to the World Health Organization (WHO), there are more than 650 million obese individuals worldwide. Since 2013, when the American Medical Association (AMA) classified a BMI

>30 as a disease, the prevalence of obesity has continued to rise. An expanding field of obesity research has revealed that complex physiological and biological factors, not simply willpower and personal choices, underlie this obesity epidemic.

Recently, the AMA also called attention to the devastating importance of attendant complications, including type 2 diabetes, liver disease, heart disease, and cancer. And it provided a call to action for more research into understanding the complex factors influencing weight gain, in the hopes of moving treatment options beyond basic dietary and exercise advice. But despite such progress, the global obesity epidemic continues to escalate as experts debate the root causes and mechanisms without coming to agreement.

Recent advances in therapies for obesity include the burgeoning use of new drugs that cause weight loss. Millions of people today are looking to lose "a few pounds," or more, by injecting needles containing the glucagon-like peptide 1 (GLP-1) agonist drugs originally approved by the FDA in 2005 as a treatment for type 2 diabetes to enhance insulin secretion, suppress glucagon release, and reduce blood glucose in patients. But early users noticed an additional benefit: they lost weight. This useful side effect drove sales into the stratosphere, and it is now anticipated that by 2035, the worldwide expenditure on GLP-1 agonist medications will hit the $100 billion mark, with the United States contributing about $70 billion to this total. Ozempic, to name one such drug, has become a well-known brand name far outside of diabetes circles.

After all that, you may be surprised to learn that no one knows for sure how GLP-1 agonists cause weight loss, but one significant line of evidence indicates GLP-1 agonists stimulate the vagus nerve. This idea prompted Sangeeta and me to wonder if we could find a way to recapitulate the drugs' effect by stimulating the vagus nerve in a different way, without the weekly injections that GLP-1 agonists entail *or* surgery. We would use a device that can be found worldwide in almost every hospital and medical clinic, not to mention every submarine: an ultrasound machine.

We established a collaboration with GE Global Research in Niskayuna, New York, and began working with medical ultrasound machines. These machines are common enough that you probably know they work by transmitting sound waves that bounce back from structures in the body to produce an image—of a fetus in utero, for example, a stone in your gallbladder, or hopefully the absence of a lump in dense breast tissue. As someone who'd spent years thinking about different ways to stimulate nerves with various forms of energy, Sangeeta knew that we could also use ultrasound to direct energy into the body in a concentrated manner toward a precise target, called *focused ultrasound*, to activate neurons. With focused ultrasound, we wondered, could we precisely stimulate the vagus nerve from outside the skin, providing therapeutic benefits without the need for surgery?

We usually feed lab mice a tiny nutritional equivalent of a healthy human salad, combining protein from fishmeal or soy, carbohydrates from corn or wheat, healthy fats from vegetable oils, with an added sprinkling of vitamins and minerals. Scientists studying obesity, however, feed their mice something more akin to a cheeseburger, fries, and chocolate shake, a diet high in saturated fat, simple sugars, and ultra-processed foods. A "Western diet" is what scientists actually call it, for shorthand, because it is nutritionally similar to what is so widely consumed in advanced economy nations. A Western diet is what we used to fatten up our mice and double their body weight.

Sangeeta and I then worked with our team to see if we could make the fat mice thinner by shining a focused ultrasound probe over their abdomens to stimulate the vagus nerve where it traversed the liver. We divided the obese mice into two groups, and every day for four months, we treated one group by applying the focused ultrasound to the vagus nerve at the liver. The others, the control group, we treated by holding the probe over the liver but without turning it on. The treatment protocol, which lasted for a total of only a few minutes each day, delivered a one-minute stimulation period, followed by a thirty-second period of rest, followed by another one-minute stimulation. After studying one hundred mice for sixteen weeks, we discovered that noninvasive focused ultrasound caused significant weight loss. But the benefits did not stop there because this ultrasound treatment also reduced the severity of obesity-associated inflammatory and metabolic derangements, including reduced levels of inflammatory cytokines and lipids and reduced *abdominal adiposity*, or belly fat.

In March of 2021, we published the study in *Scientific Reports*, concluding that "these findings suggest a previously unrecognized potential of hepatic focused ultrasound as a possible novel noninvasive approach in the context of obesity." In other words, a window into a future where focused ultrasound via a safe, noninvasive device could treat obesity. Laughing out loud, Sangeeta said, "When the clinical trial happens, I'll be the first one to get enrolled!" Well then, sign me up second.

Perhaps unsurprisingly, we are not alone in our interest in trying out this new approach. As I write, Altmetric, a company that tracks and analyzes the online activity and attention received by millions of scholarly articles and research outputs, has ranked our paper among the top 2 percent in their calculated rating system (called an *attention score*), indicating an exceptional level of interest and influence among our peers and various other sources, including news media, social media, blogs, policy documents, and online reference managers. Apparently, the paper's discoveries are not only scientifically significant but also, echoing my informal survey of colleagues at our lab meeting, highly relevant and engaging to a broader audience beyond academia. More evidence that a lot of people want to lose a few pounds, not that you were doubting.

So, what is focused ultrasound? And how does it cause weight loss? The first question is a whole lot easier to answer than the second.

The typical ultrasound machine you may have seen or even experienced for medical imaging generates high-frequency sound waves. These waves are used to produce images of structures inside the body. A handheld device, called a *transducer* or probe, contains piezoelectric crystals, which are quartz or ceramic materials that vibrate when an electric field is applied to them. Because they vibrate at very high frequencies, this produces the ultrasound waves that are then emitted from the transducer into the body. Since the body is primarily composed of water, the ultrasonic waves travel with relative ease, but as they encounter different structures, such as organs and tissues, their behavior changes due to variations in density and composition.

Organs and other bodily structures reflect, absorb, or scatter the ultrasound waves differently than water, creating echoes that travel back through the aqueous medium of the body and return to the ultrasound transducer. When these echoes hit the piezoelectric crystals in the transducer, they cause the crystals to vibrate differently, generating electrical signals proportional to the strength and frequency of the received echoes. The ultrasound machine's computer processes these signals to create an image, or *sonogram*, as a visual representation of the internal structures of the body, a technique widely used for visualizing internal organs, assessing fetal health in pregnancy, and diagnosing various conditions.

Peripheral focused ultrasound (pFUS), on the other hand, employs a different transducer specifically designed to make higher-intensity, more narrowly targeted sound waves. These deliver concentrated energy to a specific, small area of the body, enough energy to stimulate nerve activity. In contrast, imaging transducers employ lower energy levels to safely create diagnostic images without causing tissue damage or nerve stimulation. Some peripheral focused ultrasound systems also incorporate real-time imaging capabilities, a combination of imaging and therapeutic ultrasound in one system. Since the FDA first approved focused ultrasound for clinical use in 2004, it has been an option for treating uterine fibroids, pain from cancer metastasis to bone, and other conditions. Ongoing and planned clinical trials place focused ultrasound at the forefront of noninvasive nerve stimulation research for the treatment of Alzheimer's disease, substance use disorders, post-traumatic stress, and cancer.

Like implanted vagus nerve stimulators, focused ultrasound can deliver short bursts of energy to coax the vagus nerve into firing its own action potentials, but by a different mechanism. The mechanical energy of the pFUS sound waves is converted into a form of pressure or mechanical stress at the targeted nerve. This stress prompts the nerve to fire its action potentials. Thus, focused ultrasound can stimulate the nerve without causing nerve damage. Electrical stimulation and pFUS have the same net effect on the vagus nerve, prompting action potentials to travel as signals up into the brain and as signals traveling down into the body—the input and output in reflexes.

Now that we have explained how focused ultrasound can stimulate nerves, we can move on to the second, more difficult question: How does stimulating the liver's vagus nerve cause weight loss?

One reason may be that focused ultrasound activates vagus nerve fibers there that convey satiety signals to the brain. When food enters the stomach, the vagus nerve sends signals to the brain stem, which it relays to the brain centers that stimulate satiety and inhibit hunger. Enhancing these vagus nerve satiety signals induces a feeling of fullness more quickly and with less food, effectively tricking the brain into believing the stomach is fuller than it is, thereby reducing overall calorie intake and aiding in weight loss.

A second contributing factor could be the role of the gut microbiome in the development of obesity. In a recent study, thin mice became obese and consumed more food after receiving gut bacteria from obese mice. Although all the mechanisms are not yet worked out, it's clear that the metabolic activities of the bacteria in the gut influence the availability of nutrients from the food being consumed, and the resultant feelings of hunger or satiety. Introducing the microbiome from obese mice alters the gut flora of the thin mice and somehow changes the signals traveling from the gut to the brain via the vagus nerve, signals that affect appetite regulation, energy use, and even insulin sensitivity. And here's the proof: When the scientists cut off those vagus nerve signals prior to the microbiome transfer by severing the nerve below the abdomen (an intervention that is not fatal because it is below the level of the heart and lungs), the animals failed to develop the increased appetite, altered metabolism, and weight gain. In other words, there is a microbiome-gut-brain axis regulating body weight, and your brain "knows" about your gut's microbiome because of signals traveling in your vagus nerve.

A third reason vagus nerve stimulation would cause the mice to lose weight in our study stems from research into how GLP-1 agonists produce weight loss by activating vagus nerve signals to the brain. GLP-1 molecules activate cells, including neurons. This activation occurs because GLP-1 binds to another specific receptor molecule that resides on the cell surface of many vagus nerve sensory fibers. These fibers are *afferents*—meaning they provide input signals from the body to the brain. Researchers have studied the vagus nerve–dependent effects of GLP-1, experimentally interfering with the GLP-1 receptors and then administering GLP-1 to the mice. They noticed that blocking the GLP-1 receptors in the vagus nerve significantly inhibited the ability of the drug to produce satiation or a sense of fullness in response to GLP-1 injections. In other mouse experiments, investigators cut the vagus nerve below the diaphragm and again found that administration of GLP-1 agonists failed to enhance satiety.

And similar experiments have been done in humans. In one study, researchers compared two groups. The first group consisted of twenty individuals who underwent partial vagotomy surgery for treatment of ulcer disease or cancer, where the branch of the vagus nerve that communicates with the stomach is severed (cutting the vagus is called a *vagotomy*). The second group consisted of ten healthy controls. Both groups received infusions of GLP-1 or saline after consuming standardized meals, but significantly, GLP-1 reduced subsequent food intake in the controls but not in the vagotomized individuals. GLP-1 also sped up gastric emptying in controls but not in the vagotomized group, and it enhanced insulin secretion in the controls, but not in the vagotomized subjects. Thus, in mice and in people, GLP-1 injections are changing how the vagus nerve functions, modifying in ways not yet fully known reflexes that underlie body weight, food intake, and glucose and lipid metabolism.

There is much more to learn. But we do know that our lab's experiments using peripheral focused ultrasound to stimulate the vagus nerve in the liver activated pathways that, like GLP-1, decrease glucose levels, increase insulin levels, and cause weight loss. GE HealthCare has announced a major new collaboration with Novo Nordisk to move focused ultrasound closer to clinical testing. Catalyzed in part by the immense success of injectable diabetes and obesity treatments like Ozempic, Wegovy, Victoza, and Saxenda, many labs and companies are now pursuing drug-free, noninvasive approaches, including focused ultrasound, for managing body weight and type 2 diabetes.

TREATING INFLAMMATION WITH FOCUSED ULTRASOUND

After learning about obese mice and pFUS, perhaps you wondered, is it possible to use focused ultrasound to stimulate the vagus nerve to suppress inflammation? That's thinking like a physician-scientist! Recently, Sangeeta had the same question, and she embarked on experiments with human volunteers to target the spleen with focused ultrasound to see if she could inhibit inflammation.

The spleen has two main jobs. The first is to filter old and damaged blood cells from your circulation, which is important but not particularly relevant to what we wanted to do in our study. The second job is to house white blood cells, including lymphocytes that make the antibodies you need to fight infection. Other white blood cells also reside there, including monocytes and macrophages, and these are the cells that produce cytokines during infection or injury. But sometimes, as in Janice's case, the overproduction of cytokines in the spleen produces the dangerous condition called *cytokine storm*. This function of the spleen is highly relevant for studying the mechanisms of inflammation and how to reduce it.

Our colleagues in Minnesota used focused ultrasound directed toward the spleen of mice with severe arthritis and discovered that within a few days, daily treatment reduced joint swelling, improved the clinical score, and modulated the expression of inflammatory genes in the animals' white blood cells. Back in our lab, before advancing to a human study, Sangeeta began by studying mice.

Using a specialized pFUS device customized by engineers at GE, she applied it to a mouse spleen (which is only ten to fifteen millimeters long). Focused ultrasound activated the splenic nerve and inhibited the cytokine storm caused by the administration of lipopolysaccharide (the same molecule, LPS, that we had once given to sailors on shore leave to stimulate their cytokine storm). Furthermore, focused ultrasound stimulation of the mouse spleen also activated the same specific group of T-cells, or T-lymphocytes, that we had previously found were activated when the vagus nerve was stimulated in either the neck or brain. This means that pFUS on the spleen had stimulated an anti-inflammatory reflex in mice that was indistinguishable from stimulating the vagus nerve itself.

Based on these mice studies, Sangeeta decided to use focused ultrasound targeting nerves in the spleen to learn whether this suppresses cytokine production in people. The human study necessitated a new "homemade" probe, a long ethical review board approval process, and writing a new study protocol, a major collaboration between GE engineers, radiologists at Northwell Health, and our lab at Feinstein. With a modified ultrasound system and a water-filled device attached to the transducer designed to focus the ultrasound waves into the human spleen, Sangeeta led the team of researchers that ultimately recruited seventy healthy human subjects and randomized them into seven groups. The focused ultrasound power was delivered within safe exposure limits and showed no adverse effects on clinical, biochemical, or hematological (blood) parameters.

Speaking of blood and ethics, this human study, unlike the 1989 study of inflammation and depression with on-leave sailors, was designed to measure the effect of focused ultrasound on the subjects' inflammatory responses without introducing an endotoxin directly into their healthy bodies. Using a laboratory method called a *whole blood monocyte culture assay* (as we had done in the SetPoint vagus nerve stimulation study for rheumatoid arthritis), our researchers collected blood and cultured it *outside* the body. This way they could assess the responses of white blood cells in the culture to compare TNF production in blood obtained before and after focused ultrasound in the spleen. When the analysis was finished, we observed that pFUS treatment had significantly inhibited cytokine production. The suppression of cytokines by targeting the inflammatory reflex in the spleen of humans paves the way for more clinical trials using this noninvasive therapy for inflammatory disorders.

MAPPING THE VAGUS NERVE FOR PRECISE TARGETING

Our colleague Stavros Zanos is confronting head on the intricate complexity of the vagus as part of a comprehensive three-year project backed by a major grant from the National Institutes of Health. His goal: to provide a complete map of the human vagus nerve, a task that requires identifying and tracing each nerve fiber, all one hundred thousand on each side, along its winding path up and down the neck and through the body, between the brain and all the body's organs and vital systems. With such a map, we hope that therapies like focused ultrasound can be better guided to their targets.

Stavros's meticulous process starts with extracting the vagus nerve from a cadaver, a delicate task given its thinness at certain points. After labeling each branch to indicate the organs to which it is connected, the team segments the nerve into tiny pieces. They embed each piece in wax, then thinly slice them into slivers only a few hundredths of a millimeter wide. They stain these slivers with specific chemicals that reveal the neurons' internal secrets, more than a century after Ramón y Cajal, a Spanish neuroanatomist in the nineteenth century, started neuroscience on a path to microscopic revelation (and shared what he saw under his microscope with the world in hundreds of elegant hand-drawn illustrations of the nervous system). Today we visualize images of vagus nerve fibers on computer screens projecting red, green, and blue fluorescence, each color representing a different nerve fiber type and holding clues to its specific function. We have come a long way from Cajal's elegant hand-drawn illustrations of the vagus nerve, which I still admire. But I am equally astonished that the project Stavros is doing only now will be the first complete map of the human vagus nerve. It is certainly about time!

Stavros and his team had already demonstrated the potential applications of a detailed map of the vagus nerve in pigs when they created a highly selective stimulator with ten distinct contact points. Guided by the map, their device was capable of selectively controlling the pigs' breathing without impacting the function of other organs. I have no doubt that even Galen, the ancient Greek physician-scientist who famously cut the vagus nerve to silence a pig, would be thrilled to see (finally!) the functional maps of the great nerve that are emerging. I believe that secrets hidden deeply within the great nerve's neuroanatomy and reflex pathways, once fully revealed, will push the boundaries of its therapeutic potential.

Flash forward not millennia or centuries but years from now for those of you who wish to lose a few pounds or more. A future bioelectronic device may well be powered by artificial intelligence algorithms to emit focused ultrasound onto your vagus nerve to help it balance appetite, body weight, serum glucose, insulin levels, and triglycerides by targeting selected vagus nerve fibers. Or to treat inflammatory conditions without surgery. The overwhelming evidence suggests this should work, already conceivable in a portable piece of technology about the size of an English muffin.

TO INVADE OR NOT TO INVADE?

There are notable advantages to having your vagus nerve stimulator implanted into your body, because it can then operate, like the vagus nerve itself, with nary a care from you over the course of the day. Ask anyone with a heart pacemaker if they would prefer a *less* automatic treatment option! VNS implants work on the same principle, effectively maintaining the activity of your vagus nerve reflexes without any conscious effort on your part.

You don't have to remember to turn your device on and off to receive a treatment once or twice or three or four times daily. The surgery to implant the devices can be done very safely and with minimal risks, ensuring compliance in patients with serious, chronic conditions because no further intervention from the patient is required. Ask Kelly Owens, or Pero Dragoje, or Toney Kincaid.

On social media, however, and even in the medical and scientific literature on vagus nerve stimulation, commentators expound upon the perceived advantages of noninvasive devices that do not require a surgery. All too often overlooked in these discussions is the difficulty, "IRL," in compliance with using such devices to self-administer therapy. Even if each noninvasive therapy session requires only five or ten minutes of active stimulation, can you, or do you want to, interrupt your busy day, every day, to stop what you're doing, set up the device, and run your own VNS therapy session? The real-world evidence indicates the answer is no.

Patient compliance with self-administered therapies is abysmal, studies show. Statistics vary depending on the specific condition and population, but it's not uncommon to see noncompliance rates of 50 percent or higher in chronic therapies, a failure rate that can contribute to worsening of the condition, increased healthcare costs, and higher rates of hospitalization and mortality.

There's always a trade-off on compliance and effectiveness when devices are noninvasive. So, rather than subdivide our discussion of vagus nerve stimulation into invasive versus noninvasive, let's approach it from another perspective, one based on neuroscience. Let's begin with two simple but revealing questions. First, does the neuroscientific evidence indicate that the device stimulates the vagus nerve? And second, does the clinical evidence indicate that the noninvasive intervention, which may or may not stimulate the vagus nerve, offers an effective treatment as evidenced by randomized controlled clinical trials?

The answers to these two questions can help you make an informed decision about whether and how to use noninvasive methods of nerve stimulation. Consider focused ultrasound because there is overwhelming evidence that it directly stimulates the vagus nerve by well-understood mechanisms that are validated by abundant experimental and clinical neuroscientific evidence. While noninvasive and promising, it is a relatively new approach. Clinical trial testing of focused ultrasound to stimulate the vagus nerve as a therapy for obesity, diabetes, arthritis, and perhaps other medical conditions is still in its early days. For now, this therapy is not broadly available for clinical use until additional trials are completed.

By contrast, you can find millions (literally millions) of recommendations on social media urging you to try something called a *transcutaneous electrical nerve stimulator*, or *TENS* unit, that you apply to the skin of your ear to noninvasively stimulate your vagus nerve. And you can find dozens of inexpensive versions of this device over the counter or online, available without a prescription, with some uses already approved by the FDA. What is the direct scientific and clinical evidence that these devices actually stimulate your vagus nerve? What is the evidence that they are effective, and if so, what benefits do they provide? These are questions my colleagues and I are asking. In the next chapter, I've shared the answers I can give you today.

8

The Ear-Brain-Body Connection

Over-the-Counter Devices with Many Potential Benefits

I thought of my whole body as an ear.
— MAYA ANGELOU

For forty years, my dear friend Ulf Andersson and I have been dreaming up experiments together, writing papers, and sharing new ideas, that is, when this brilliant and ebullient Swede wasn't teaching my then-young daughters how to fake out the goalie during a penalty kick, or guiding me on a fifty-mile skating tour of the frozen Stockholm archipelago. Now seventy-five years old and retired from his job as a professor of pediatric rheumatology at the Karolinska Institute, Ulf feels as creative and energetic on his skates as ever. But some twenty years ago, he became quite ill and was diagnosed with a large tumor on his pancreas.

Fortunately, it was benign. Unfortunately, it was obstructing the flow of pancreatic fluids and bile, so it had to be removed.

Ulf was subjected to a major abdominal surgery, a so-called *Whipple procedure*, to remove the tumor, along with the gallbladder, bile duct, and portions of the pancreas and small intestine. This left him besieged with chronic inflammation in his liver and the remaining portion of his pancreas, a "highly unwanted" but not unexpected result of the operation, as he puts it with characteristic wryness. A Whipple procedure leaves intestinal bacteria with free access to reach the remaining bile and pancreatic ducts, sites that are normally sterile. When bacteria grow in these sites, it causes a severe infection. And where you find severe infection, you find severe inflammation.

Every few months, two to four times every year, my friend was disabled by four or five days of severe abdominal pain followed by weeks of severe fatigue, reduced cognitive capacity, and, worse still, the onset of generalized anxiety and depression.

"The cure for the acute thing was easy," according to Ulf, and he avoided hospitalization from the infections by treating himself with large doses of antibiotics. Less easy were the long periods of fatigue that accumulated after each of these flares. "And then came something that I had never experienced before, generalized anxiety and depression, and that was really the horrible part. The abdominal thing is easy to cope with, but generalized anxiety was hell on Earth." This miserable cycle persisted for eight years, making him dependent on continuous psychiatric support, including daily antianxiety medications. The doses escalated each year.

In an unusual and moving convergence of doctor and patient in the same person, Ulf the physician-scientist was inspired by

Ulf the patient's experience to write up his own clinical report, with references to the scientific literature. In his career, he studied how the nervous system regulates inflammation, then he fell ill with a severe chronic inflammatory condition. He was an expert on what ailed him.

"Lessons that the patient learnt from this long clinical and scientific voyage," the document explains, "starting long before the presentation of inflammatory disease, inspired the patient to write this case report." It recalls the years before he got better:

> [Large doses of antibiotics and psychiatric medications] enabled the patient to work effectively in his profession, but he felt that his wings were cut, since he was always tired and did not have access to his full mental resources. Family life suffered badly because of fatigue and subsequent change of life style. The patient has four children, two of whom were rather young then, and it was very hard to manage to act as a vital father and a good husband. The marriage ended with a divorce, connected to these health problems.

Then, my colleague and friend, the doctor and patient, had a minor heart attack.

Soon after that, in 2017, Ulf began to use a transcutaneous electrical nerve stimulator (TENS), placing an electrode on his external ear a couple of times a day in the area called the *cymba concha*. Pronounced SIMba KONcha, the name stems from the Latin words for "boat" (*cymba*) and "shell" (*concha*), referring to the bowl-shaped cartilage surrounding the opening to your ear. This just happens to be the only place in the body where a branch of the vagus nerve comes to the skin. So, Ulf was using his TENS device as a method of so-called *transcutaneous auricular vagus nerve stimulation* (taVNS). He hoped it would ameliorate his persisting chronic abdominal inflammation, and maybe his depression.

Ulf stimulated his left ear this way for five minutes, twice a day, with a pulse width of two hundred microseconds and a frequency of twenty Hz, while varying the pulse amplitude between two and four milliamperes (mA) as tolerated. And he began to improve within two weeks. He attributes his improvement to the release of the neurotransmitter acetylcholine—*vagus stuff*—prompted by electrical stimulation of his ear in just the right spot, where he believes he is stimulating the auricular branch of his vagus nerve. Acetylcholine, that first rest-and-digest neurotransmitter to be discovered, prevents the release of the inflammatory molecules TNF and HMGB1, the latter of which Ulf believes was driving his chronic inflammation.

Now, after seven years of regular TENS use, the patient is in excellent physical and mental shape:

> *This treatment has been a success and has improved quality of life to what the patient experienced twenty years ago before falling ill. . . . He walks at a good pace in the forest for at least 1 hour every day. . . . There has been no need for psychotropic drugs throughout the last five years.*
>
> *The patient has regularly visited a cardiologist after his coronary infarction in 2013, and the outcomes of these yearly examinations with ECG at rest and during exercise are free from pathological findings.*

And when the ice is thick and not covered in snow, he skates for miles on the frozen archipelago.

AURICULOTHERAPY

Have you ever put a Q-tip in your ear, or perhaps your child's ear, and triggered coughing? This is a reflex, known as the *auricular cough reflex*, that happens because your ear is connected to your brain through a sensory branch of your vagus nerve. Normally, a cough reflex occurs as a mechanism to clear your respiratory tract of irritants. But in this case, the stimulation of the ear canal is misinterpreted as an irritant in the throat. Also called the *Arnold nerve reflex* (after Friedrich Arnold, the German anatomist and physiologist who identified it), this reflex can vary in sensitivity among individuals and is not universally experienced, but it is direct evidence that stimulating nerves in your ear can send signals to your vagus nerve and brain.

As the only part of your body where your great nerve extends a peripheral sensory branch to the skin and cartilage, your external ear is unique. Known as the *auricular branch of the vagus nerve* (ABVN or *Arnold's nerve* for short), this afferent nerve travels from the cymba concha into your brain stem *nucleus tractus solitarius* (*NTS*), the destination of other sensory vagus nerve fibers traveling from your body's organs. Since the NTS is like a router that processes incoming signals and directs outgoing reflexes, this means signals from your external ear can travel to connect to other sensory vagus nerve networks in your brain.

What social media influencers, health gurus, and clinical investigators in my lab and elsewhere usually call *transcutaneous auricular vagus nerve stimulation* (taVNS) originates from the practice of acupuncture, an integral part of traditional Chinese medicine. Acupuncture of the ear gets its first mention in Chinese texts penned by anonymous authors sometime between 770 and

221 BC, when they proposed applying a hot instrument to burn or cauterize the skin at certain points on the outer ear to cure aliments in other body parts. With the passage of time, when the instruments and equipment improved, acupuncturists swapped the dangerous hot cautery devices for small needles inserted into the skin. These needles could also be electrified.

Notwithstanding the popularly held notion that the ancients had it all figured out, auricular acupuncture (*auricular* being a word borrowed from Latin meaning "of the ear") did not gain widespread adoption until relatively recently, with the help of detailed acupuncture maps developed in the 1950s by French physician Paul Nogier. Dr. Nogier launched the new field of *auriculomedicine* after observing two patients who had scars on their ears and a history of sciatica and back pain. Both had been treated by a folk healer in Corsica whose method, like the ancient Chinese recommendation, involved burning a small area on the external ear using a hot cauterizing probe to alleviate pain elsewhere in their body.

Nogier sought an explanation. He first crafted his own version of a pressure probe by removing the ink cartridge from a ballpoint pen. With his new tool, he systematically examined patients and correlated pain arising from specific body parts with points on the ear he identified as having increased sensitivity to pressure from the end of his ear-pen device. After several years, believing he had successfully linked spots on the ear to corresponding bodily dysfunctions, Nogier presented his maps at the 1956 inaugural Congress of the Mediterranean Society of Acupuncture in Marseille. He proposed a new theory of auricular medicine. News of Nogier's theory soon spread back to China where, in 1958, a study using his ear map in two thousand patients concluded that illness in specific body parts is correlated to the points he labeled on the ear. Later publication of his auricular maps in a widely cited Chinese medical journal cemented him as a global leader in ear acupuncture.

TaVNS is based on the idea that we can safely and cheaply hack the ear to stimulate vagus nerve networks. In 2001, one of the earliest clinical studies demonstrated the basic principles for electrically stimulating the ear to modulate signals in the vagus nerve. It studied eighteen patients with acupuncture needles inserted into the skin at a painless depth of only 0.1–0.3 millimeters in the cymba concha. Pulsating bursts of electric current went through the needles for fifteen minutes daily for ten days, which decreased the patients' heart rate and blood pressure. The researchers concluded they had electrically stimulated the vagus nerve. This study ignited the imagination of countless other investigators and spawned a new industry of clinical research and medical technology based on taVNS. For example, Harvard Medical School recently studied twenty hypertensive subjects treated with taVNS using a range of specific electrical stimulating conditions. The researchers observed significant decreases in heart rate and blood pressure after five sessions delivering thirty minutes of electrical stimulation to the ear.

There are piles of clinical and laboratory studies along these lines, generating a tremendous amount of interest. But these studies also raise important questions, some of which we can answer and some not:

What are the electrical stimulating devices involved and how do they work?

What is the evidence that electrical signals in the ear are activating responses in the brain and body?

And is there any proven clinical benefit for doing this? Let's take these one at a time.

TENS

Transcutaneous electrical nerve stimulation (TENS) units are portable, battery-operated pain management tools. The origins of TENS, like acupuncture, date back at least two thousand years. A physician in ancient Rome, Scribonius Largus, recommended that his gout patients stand on electric torpedo fish to experience pain relief from their natural electrical discharges (giving us, if not an entirely practical analgesic solution, certainly quite an image).

Much more recently, in 1965, a Canadian psychologist and a British neuroscientist proposed a theory of pain perception that would explain the therapy Largus touted long ago. In Ronald Melzack and Patrick Wall's "gate control theory," pain is modulated by a "gate" in the spinal cord. Depending on whether the gate is open or closed, it either blocks or allows pain signals to travel up to the brain. The therapeutic insight derived from this theory is that nonpainful inputs to the spinal cord, such as electricity, can close the gate, thereby preventing pain sensation. This theory spurred the development of TENS units, which deliver small electrical impulses through electrodes placed on the skin, closing the gate, live fish not required.

Most TENS units are compact, nine-volt-battery-powered devices that generate adjustable bursts of electricity. As with the implantable VNS devices and as Ulf Andersson detailed about his self-treatment, variable parameters include the current's strength (amplitude), pulse width, and frequency. Electrodes, typically adhesive pads, are placed on the skin. One electrode acts as the ground, while the other delivers the therapeutic pulses.

In 1976, the FDA classified TENS units as Class III medical devices, subject to the most stringent regulatory control. However, widespread safe use led to their reclassification as Class II devices, and they are now available over the counter for twenty-five to one hundred dollars. TENS units are used at home and in clinics for various pain conditions, though effectiveness varies. The FDA advises consulting healthcare providers before use, especially for those with underlying medical conditions or implanted devices.

Applying a TENS unit to the ear is a method of taVNS. Depending upon the location of the electrodes and the specific neuroanatomy of the subject, using a TENS device on your ear may deliver electrical pulses to the auricular branch of your vagus nerve. Various companies market a variety of handheld and clip-on electrodes fitting the ear. (At his kitchen table, Ulf invented a headset apparatus to hold his TENS unit in place.) Typically, the technique involves one or more intermittent stimulation sessions daily, each lasting from five minutes to an hour or more, for a total daily stimulation time ranging from five minutes to five hours. Ulf maintains his results with two five-minute sessions a day.

The effects of various types of taVNS have been studied in hundreds of clinical trials, with exciting results demonstrating significant improvements in epilepsy, depression, anxiety, pain, opioid withdrawal, arthritis, inflammatory bowel disease, diabetes, headache, and more. Researchers have proposed many hypotheses to explain these benefits, reasoning that the electrical pulses selectively stimulate vagus nerve endings in the ear. But in reality, these findings raise as many new questions as answers.

To begin with, it's important to distinguish between different types of vagus nerve stimulation. Or, more precisely, we must understand what vagus nerve fibers we are stimulating, of the two hundred thousand or more that are available. And while blogs, social media posts, and self-help websites recommend you "stimulate your vagus nerve" by "stimulating your auricular branch of your vagus nerve," words matter here. We need to be careful not to gloss over the differences between surgically placing an electrode directly onto the vagus nerve trunk on the left side of your neck, home to one hundred thousand vagus nerve fibers, or directing focused ultrasound at the vagus nerve in the liver or spleen, as compared to stimulating a tiny branch of the vagus nerve comprised of only a few hundred sensory fibers beginning at your ear.

In addition, the innervation of ears varies between individuals, so we cannot be sure whether your TENS unit is stimulating your ABVN specifically or one of the other nerves in your ear. In and around your ear and cymba concha there are small branches of the trigeminal nerve (cranial nerve V), the facial nerve (cranial nerve VII), the cervical plexus, and lesser occipital nerves, as well as the vagus nerve (cranial nerve X). When you apply a TENS unit electrode to your ear, the electric current may travel across a stimulation field on the skin and stimulate several of these nerves. So, while we can say for certain that using a TENS device on your ear is "electrical nerve stimulation through the skin," and Ulf can say that after he started using a TENS device on his ear, he got his health and his life back, we don't know that these effects are from *vagus* nerve stimulation, ABVN stimulation, or some other ear nerve stimulation.

Regardless, all the neurons in your ear, body, and brain are ultimately interconnected in one way or another through vast, overlapping neural networks. When we send signals into the brain stem via the ear's sensory nerves, it is possible the brain neurons receiving these signals, in turn, activate reflex signals traveling back to the body in the vagus nerve, or in other nerves. Therefore, we are left with the all-too-real scenario that the brain is pumping out signals through so many different nerves that no one can say for sure what is happening. Saying that a specific bodily response to an electrical burst through an ear electrode is the same as "vagus nerve stimulation" produced when we implant an electrode directly onto the vagus nerve in the neck is, at best, an imprecise description. At worst, it is completely incorrect.

Yusuf Ozgur Cakmak at the University of Otago in Dunedin, New Zealand, recommends that perhaps we should refer to this therapy as *auricular electrical stimulation*, or *AES*, since it describes what we think is happening more accurately. The broader *transcutaneous electrical nerve stimulation*, or *TENS*, is also correct. At any rate, if you see claims about the benefits of "taVNS," you'll now be armed with a better understanding about the limits of current science and take these claims with a grain of salt. But while scientists don't understand how electrical bursts to the ear are processed by the brain to in turn produce specific responses in the body, that doesn't mean you cannot benefit from them.

NEW FRONTIERS IN TENS RESEARCH AND PROVEN BENEFITS

One day in the lab, Sangeeta and I were discussing this very problem of the mechanisms and effects of ear stimulation. We Googled "ancient Chinese acupuncture map of the ear" (not knowing it was the much younger brainchild of a doctor born in France). On the cymba concha, near the opening to the external auditory meatus (the passage into your ear), was a point clearly labeled SPLEEN. Spleen, that mysterious, bloody, squishy organ of the immune system and participant in cytokine storm. What if stimulating this point with a TENS unit could suppress our cytokine responses?

We decided to donate our own blood to this cause.

We knew from our studies of implanted vagus nerve stimulators and focused ultrasound in people that stimulating the vagus and splenic nerves suppresses the production of cytokines. And from our studies in mice, we also knew that vagus nerve stimulation for just five minutes produced a lasting effect that inhibited cytokine production for up to twelve hours afterward. So, we thought, maybe a five-minute TENS session in our own ears would suppress cytokine release for several hours? Maybe that is why TENS at the ear confers some benefits in some clinical trials of inflammatory conditions?

Five of us gathered around the conference table in my office. We each donated a few tubes of blood, then used the TENS unit on our left ear for five minutes. For me, at first, the buzzing felt like a gnat on my skin, but it didn't hurt. After a few minutes, I felt "butterflies in my stomach," a sensation that I found reassuring. When that happened, I thought just maybe the electric bursts really were hitting my vagus nerve, which reaches all the way into my stomach and other abdominal organs. But who knows? Others described the feeling as having a facial, which is also interesting. The auricular branch of the vagus nerve has many interconnections with the trigeminal and other nerves innervating the throat, neck, and face.

Two hours later, we donated some blood again, which Sangeeta analyzed for cytokine production. The results were astonishing: Four out of the five of us had between 50 and 75 percent decreases in our cytokine production. It was one of those days. Holy shit.

We measured cytokine production using the same methods we had used to test the effect of vagus nerve stimulation in clinical trials of vagus nerve stimulation implants. Although this did not prove that the vagus nerve was responsible, we were certainly intrigued. It's a good adage in science and medicine that when you hear hoofbeats in New York City, think of horses, not zebras. So even though we did not have a direct measure of the TENS unit stimulating our vagus nerves, it seemed a reasonable explanation to begin with.

However, our lab colleague "John" did not experience a significant reduction in cytokine levels from the TENS unit on his ear. Interesting. As we pondered this, we realized that after John's ear was stimulated, he jumped up and ran from the room, saying, "I have to pee!" John is normally modest and reserved, so this behavior was strange. We wondered if there was any connection between his bathroom urgency and his normal cytokine responses despite having received the same TENS stimulation as the rest of us.

That was when Sangeeta, having turned to the ear acupuncture map on the computer screen, said, "Oh wow, Kevin, you have to see this!"

On the map right next to SPLEEN, another point was prominently labeled BLADDER.

Had the TENS stimulated John's bladder instead of his spleen?

Four days later, we repeated the experiment, this time being even more careful to position the TENS unit probe onto the spleen point, according to Nogier's map, and avoid the bladder point. Two things happened next: one that did occur, and one that did not. What did happen is John's ear stimulation decreased his cytokine production this time. What did not happen is that he did not run out to the bathroom.

Had we now proved that stimulating the ear with a TENS unit stimulates the vagus nerve inflammatory reflex? Still no, but it was certainly intriguing. Especially when considered in light of hundreds of other clinical results that have addressed whether electrical signals in the ear affect the vagus nerve, brain, and body.

Neuroscientists and clinical researchers have used functional MRI (fMRI) to study whether electrical stimulation of the external ear affects the activity of brain networks. These results reveal some overlap in the brain regions that are activated by both methods. In a controlled study using electrodes to stimulate the cymba concha, fMRI images showed enhanced activity in the brain regions associated with vagus nerve input and output centers, including the NTS, the dorsal raphe, locus coeruleus, and nucleus accumbens. In another study of the relationship between ear stimulation and brain responses, researchers enlisted a method of electroencephalographic testing called *far-field evoked brain stem potentials*, a type of electrical signal recorded from electrodes placed on the scalp. Analysis of these brain stem potentials in volunteers receiving ear stimulation suggests that taVNS does activate vagus nerve nuclei in the brain stem.

Healthy volunteers receiving taVNS showed dilated pupils and EEG activity, suggesting that TENS may improve our ability to pay attention. This enhancement has been attributed to changes in the brain neurotransmitter gamma-aminobutyric acid (GABA), an inhibitory neurotransmitter that plays a role in regulating anxiety and other brain functions during vagus nerve stimulation. Yet another study in healthy subjects found that ear stimulation increases cerebral blood flow to enhance levels of oxygen in the prefrontal cortex. Additionally, a series of clinical trials using ear TENS in epilepsy patients revealed that it reduced seizure frequency, intensity, and duration, and in some patients even reduced the need for antiepileptic drugs, similar to surgically implanted VNS.

There are too many studies to mention them all here. This large body of data—though it doesn't prove a cause-and-effect relationship—does point to a possible role for vagus nerve reflexes in the brain that are activated by TENS electrodes applied to the right spot on the ear. Unfortunately, these methods have not been standardized to the point that we know where this "right spot" is.

Here's what I can say with certainty: Taken together, it seems hundreds of available laboratory and clinical studies provide evidence that stimulating the ear with a TENS unit can cause changes in the cardiovascular and immune systems of humans. Many of these changes are similar to the effects of direct vagus nerve stimulation in people and in animals. However, we still lack direct proof in humans.

The body's physiological responses to any intervention, even ear stimulation, are the result of complex interactions between many systems, including parasympathetic, sympathetic, endocrine, and vascular. We don't have a good way to directly measure your vagus nerve activity with an electrophysiological probe or any other method, so we cannot ascertain a direct relationship between the electrical bursts in your ear and the action potentials in your vagus nerve.

For now, it seems we have hit the limits of what is knowable about ear stimulation mechanisms. Nonetheless, despite not knowing how it works, there are lots of interesting results from clinical trials. And many important findings indicate taVNS can have clinical benefit. Let's look at a few.

Inflammation

My colleagues and I have continued to explore the use of ear TENS in treating other conditions, and we have observed inhibition of cytokine production in healthy volunteers and in patients with rheumatoid arthritis, lupus arthritis, and other conditions. In a series of peer-reviewed clinical trials, we have also reported that ear TENS can improve the symptoms of pediatric Crohn's disease, pediatric nephrotic syndrome, and adult rheumatoid arthritis and lupus arthritis. Although the evidence for clinical responses is compelling, these trials are clinical experiments, not the final answer. The number of patients we have studied to date is relatively small. Larger, randomized sham controlled clinical trials are now necessary before we can make broad recommendations.

Opioid Withdrawal

The FDA has approved the commercial sale and use of ear devices for the treatment of opioid withdrawal syndrome. These are available by prescription for patients suffering from opioid withdrawal to reduce the duration and severity of joint pain, sweating, insomnia, and other withdrawal symptoms. These TENS-like devices electrically stimulate several nerve branches to the ear, including the trigeminal, facial, glossopharyngeal, vagus, and occipital nerves, and have few side effects.

Headache

The FDA has also granted approval for a device that is held against the skin of the neck to deliver electrical bursts in the vicinity of the vagus nerve as a prescription therapy for some types of cluster and migraine headache in adult patients. Because this TENS-like device stimulates superficial nerves in the neck, it can also cause contraction of the platysma muscle, which accounts for the side effect of lip curling.

Cardiovascular Health

Several, but not all, clinical studies have shown it is possible to use ear TENS to lower heart rate and blood pressure and to increase heart rate variability (HRV). For example, a 2022 study of healthy volunteers systematically tested ear stimulation and observed significant enhancement of "vagal tone" as evidenced by increases in the high-frequency range of the HRV spectrum, noted for its dependence on vagus nerve activity (and these improvements were better in volunteers who entered the study with a poorer basal level of vagal tone).

After his heart attack and subsequent daily TENS unit intervention, Ulf used a Vagus ECG Smartwatch, developed by Vagus Health in Cambridge, UK, to monitor his heart rate, heart rate variability (HRV), and synchronization between HRV and controlled breathing (respiratory sinus arrhythmia or RSA, and RSA synchronization, i.e., the RSA Sync measure). His smartwatch let him compare data from before, during, and after his TENS sessions. After gathering and charting this data regularly over a six-month period, Ulf says the results clearly indicated "improved synchronization of cardiac and respiratory activities." The results of larger, randomized clinical trials are needed to interpret and confirm these findings.

Anxiety and Depression

Because surgical implantation of vagus nerve stimulators is approved by the FDA for treating anxiety and depression, many research centers are pursuing whether ear TENS (or AES) might also be effective in this condition. In one study, performed as a collaboration between Harvard Medical School, in the U.S., and the China Academy of Chinese Medical Sciences, in China, Dr. Peijing Rong and colleagues treated ninety-one patients with taVNS for twelve weeks. They observed a greater clinical improvement as assessed using a standardized Hamilton Depression Rating Scale as compared to control patients receiving a sham therapy. A review of twelve clinical studies that included 838 participants also reveals taVNS significantly improves Hamilton Depression Scale scores and provided evidence that the clinical response rates to ear intervention were comparable to the clinical effects of pharmacological antidepressants.

Ulf is officially retired as a professor at the Karolinska Institute, but his research continues, now focused on different applications for the headset-style taVNS unit he built, literally at his kitchen table. It has aluminum electrodes (and thus no need for messy electrode gel), and it's designed to stay comfortably in place with the appropriate amount of pressure, "not too little and not too much," Ulf notes.

He has spent the past year completing paperwork to clear the regulatory hurdles for one study he is planning, of 128 patients with rheumatoid arthritis at seven different hospitals in Sweden. Half of the subjects will receive a placebo treatment, and the other half will receive electrical stimulation at the cymba concha daily for three months. Ulf told me that the hospitals are eager to get started.

He is also planning a smaller study that is close to his pediatrician's heart, of fifteen children with Duchenne muscular dystrophy. These children are healthy at birth, but due to a genetic disorder, their skeletal muscles weaken to the point where, by the age of six or seven, they need a wheelchair. They also develop severe inflammation, often treated with high doses of corticosteroids. The steroids have especially bad side effects in a growing child, including osteoporosis that can lead to painful fractures in the spine and other bones.

For the planned study, families will learn how to use Ulf's TENS unit on their child. Then they will do this at home for a week, with blood samples taken before and after to see if the treatment reduces serum cytokine levels. He is optimistic. "If we can see that this works, then of course we will extend the time," he says. "If so, that could be a chance to reduce the requirement for corticosteroids, and that would be a very good deed."

"I was always very afraid of the day that I would be forced into retirement," Ulf confides. In Sweden he had no choice at the age of sixty-five. (Now it is sixty-nine.) Ten years later, he says, "My head is better now than when I was fifty-five." As detailed in his case report, he attributes this in no small part to the twice-daily use of his auricular stimulation device. "It has always been my intention that science is fun but that the meaning of science is to be of benefit to sick people. With the taVNS device, it may finally happen before I push up"—he pauses to find the English idiom he's looking for—"daisies."

Much more study is needed to determine the most beneficial or effective stimulation parameters for TENS units, including intensity, pulse width, waveform, and frequency. Until now, most clinical trials have used only a few selected variables in these parameters, leaving open questions about which of the countless possible settings are best. Currently, we do not know how to recommend an optimal or personalized setting for an individual patient, or how to tailor the stimulation session to a specific clinical condition. And since nerve locations can vary somewhat from person to person, we don't know how to place the electrodes on the ear to hit the "right spot" on a given person.

We also know very little about the duration of the effects of taVNS, and very little about the optimal frequency of use. Is it better to do it once, twice, or three times a day, and for how long? And how long does the post-stimulation benefit, if any, persist? One study of insomnia patients observed improvements in sleep persisting up to three months after a two-week period of twice-daily taVNS therapy.

Sangeeta and I also reported that taVNS inhibited the production of some cytokines (but not all) for up to twenty-four hours after a five-minute taVNS session in healthy volunteers. And we found in rheumatoid arthritis patients that taVNS alleviated pain and swelling for several days following treatment. So, while there is some evidence taVNS can have lasting effects beyond the treatment period, we await and look forward to more comprehensive and long-term follow-up studies.

Clearly, we still have much to learn about the clinical effects and underlying neurological mechanisms of transcutaneous electrical nerve stimulation methods. But the safety, low price, and potential useability drive high interest as part of the expanding trend for wearable devices and at-home, self-administered care. As popular interest expands and intensifies, I look forward to the time when the science and clinical studies catch up to the popular demand. More studies will clarify the facts, reduce the confusion, and optimize potential benefits from placing a TENS unit (or maybe even a Q-tip?) on your ear.

I am glad my friend Ulf is working on this, along with many others. And I am gladder still that he is feeling better.

III

Great Expectations

Everyday Tools with Promise

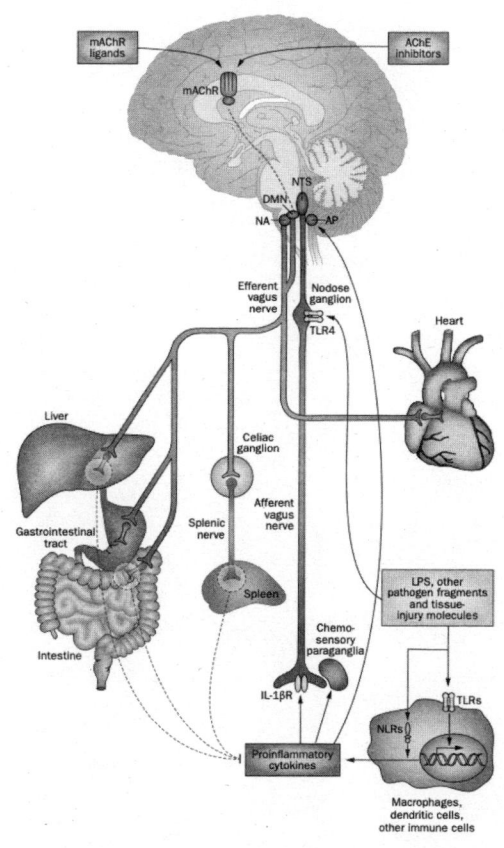

9

Meditation and Breathwork

Live long and prosper.

— SPOCK

I was a child, and it was the dawning of the Age of Aquarius. Or it was supposed to be. My growing-up years spanned the 1960s and '70s, that transformative period in America famously marked by a cultural revolution of social and political upheaval: the civil rights movement, protests against the Vietnam War, and sexual liberation. This turbulence also powered the imagining of new futures and technologies arising out of the chaos and change.

Everywhere I looked, in my classrooms, in the news, and in movies and TV, I was inundated with bright promises of space exploration, the early computer industry, and otherwise unparalleled progress in scientific and technological discovery. I was smitten with *Star Trek* and *Dr. Kildare*, shows about seemingly predestined opportunities to explore the unknowns of the cosmos, on the one hand, and on the other, the microscopic universe inside our own cells. In keeping with this fervor, in fact, the National Institutes of Health expanded during this time and the research enterprises of the pharmaceutical and healthcare industries grew exponentially. Concepts once relegated to science fiction emerged as tangible possibilities.

The era encouraged spiritual exploration, too, with Western folks drawing from Far Eastern religions like Buddhism and Hinduism. A growing interest in meditation and breathwork turned spiritual guidebooks like *Zen Mind, Beginner's Mind* by Shunryu Suzuki into Western bestsellers. And along with this spiritual and mental health awakening came a renewed interest in physical well-being. Jogging, aerobic exercise, and yoga exploded into the mainstream, reflecting a broader cultural shift toward health consciousness. TV and big-screen celebrities like Jack LaLanne in the 1960s, Arnold Schwarzenegger in the 1970s, and Jane Fonda in the 1980s motivated millions to pursue new exercise routines. Tom Hanks embodied this ethos as the character Forrest Gump, when he started running across America and kept going for 1170 days (and 16 hours).

These late-twentieth-century promises of better, more healthful living through meditation, exercise, and other lifestyle changes prompted early skeptics and believers alike to raise legitimate questions. Are these practices efficacious in improving health and well-being? Do the results justify the investments of time and money they require? And what is the scientific evidence of mechanisms by which these practices improve the brain and body?

Today, my lab's work using new technologies to explore and tweak the nervous system is converging on these same questions. Growing evidence is beginning to illuminate the role of the vagus nerve in the positive outcomes associated with meditation and breathwork, as well as cold exposure and exercise, which I will discuss more in this chapter and the next. Let's consider these activities one at a time to see how you might stimulate the protective and healing reflexes of your vagus nerve at home and in your everyday life.

MEDITATION

On a fall day in 2006 at the Menla Center in Phoenicia, New York, I had the honor of sharing the stage with Elizabeth Blackburn, a Nobel laureate from UCSF, and Tenzin Gyatso, the XIVth Dalai Lama, before an audience of three hundred monks and scientists. A sanctuary dedicated to Tibetan healing therapies, wellness, and spirituality, Menla operates under the auspices of Tibet House US, one of several nonprofit organizations around the world established at the request of the Dalai Lama to safeguard and celebrate Tibetan culture. Nestled in a glacial valley encircled by vast forests and the Catskill Mountains, Menla is an idyllic setting for a three-day meeting about "Longevity and Optimal Health: Integrating Eastern and Western Perspectives."

The event, a collaboration between the Columbia University Integrative Medicine Program and Tibet House, aimed to bridge the gap between leading scientists and Buddhist practitioners on the science of health and aging and the ancient Indo-Tibetan practices of meditation, yoga, and diet. The Dalai Lama, who has authored several bestsellers on the interplay between mental and physical health, champions emotional well-being and the power of meditation to foster a serene, focused mind. I was eager to present my findings on the inflammatory reflex and the potential role of the vagus nerve in the health benefits of meditation.

Elizabeth and I alternated in briefing the Dalai Lama on the scientific insights from the previous day's sessions. We covered a broad array of topics ranging from the neuroscience of meditation to Elizabeth's discovery of telomeres, the work that won her the Nobel Prize.

Our host, His Holiness, has a long fascination with science and technology. A lifelong learner, he has expressed willingness to adapt or modify Buddhist teachings as necessary if scientific evidence reveals the beliefs are incorrect. When he asked me about the vagus nerve during our discussion onstage, I recognized it was a testament to this open, curious mindset. He is a spiritual person firmly based in reality.

I explained how a vagus nerve reflex inhibits inflammation. From across the stage, the Dalai Lama seemed to peer right into my thoughts, then inquired with a kind and frankly knowing stare, "About this vagus nerve, is there one or two of them?"

"There are two," I said, "one on each side of the neck. They run down through the chest and into the abdomen, where they touch all the body's major organs along the way."

"Are they in the front or back?" the Dalai Lama wanted to know.

When I told him they were in the front, he smiled and nodded as if to say, "That's what I thought, thanks!"

Later, as I stood at the edge of the stage chatting with a small group who had gathered after the Dalai Lama's departure, a monk in orange robes asked if I knew why His Holiness had taken that line of questioning. He went on to explain an ancient Tibetan-Buddhist meditative practice in which the meditator envisions a cloud of blue light or energy around the top of their head, which they then channel as two beams running down the front of their neck, one on each side, continuing into their chest and abdomen.

"Cool," I said, smiling. The monk smiled back and said quietly, nodding, "Yes, very cool. Very cool."

From paying attention to every breath or repeating a word or mantra to sitting cross-legged while channeling blue light down both sides of your neck or walking a sandy shoreline, meditation encompasses a range of practices and techniques designed to nurture inner peace, compassion, and the ability to recognize the fleeting experiences of your sensations, thoughts, and emotions. Meditation is widely practiced to promote mental clarity, emotional stability, and adaptability to change.

There are many different methods of meditation from many different traditions, and they are best detailed in other books on the subject. I am not a meditation expert, but I do meditate. I assume a relaxed physical posture and hold attention on my breath or on a specific intention, such as wishing for the well-being of others. Then I simply note the emotions, thoughts, and sensations that arise in the present moment, practicing a state of nonreactive awareness of their appearance and disappearance. Hundreds of clinical studies using brain imaging with fMRI and PET scans have attempted to determine what these meditative practices are doing to the brain, nervous system, and vagus nerve. Reading these studies, I can't help but wonder: What is happening inside my own brain when I meditate? The results so far, while interesting, leave plenty of unanswered questions.

For example, one systematic review of 78 of these functional neuroimaging studies, including a total of 527 participants, revealed consistent patterns of brain activation and deactivation in subjects doing various meditation practices; each exhibited a different pattern of brain activity on fMRI and PET scans. Specific brain regions, such as the insula, prefrontal and supplementary motor cortices, anterior cingulate cortex, and frontopolar cortex, were consistently activated across multiple meditation techniques. However, the brain activity patterns observed across different meditative practices were largely distinct. After reviewing all this data, the authors concluded the jury is still out, and "it is imperative that neuroimaging investigations of meditation begin to rigorously relate findings to behavioral and self-reported outcomes, in order to understand how (and to what extent) these changes in brain activity are related to actual cognitive-affective change and benefits in practitioners." In other words, much more research is needed for us to understand how meditation works in the brain to influence psychological and physical well-being.

Hundreds of other clinical studies have explored how meditation influences cardiovascular health. As explained in chapter 2, vagal tone, a measure of parasympathetic nervous system activity, is calculated from changes in the time in between individual heartbeats, which vary through respiratory inhalation and exhalation. Vagal tone is increased when the parasympathetic activity is increased, which promotes relaxation and utilizes acetylcholine to decelerate your heart rate. In some studies, meditation practices enhance the meditator's vagal tone, with measurable effects including slower resting heart rate and reduced blood pressure. Based on these results, some researchers have suggested that meditation acts as a "vagus nerve stimulator."

But other studies have failed to reach this same conclusion. A review of nineteen randomized clinical trials that included more than one thousand subjects indicated that meditation did not increase vagal tone. According to these authors, "There is currently insufficient evidence to indicate that MBIs [mindfulness-based interventions, one type of meditative practice] lead to improvements in vagally mediated HRV over control conditions."

Given the public health implications, the American Heart Association sponsored a study to define recommendations for the potential advantages of meditation to protect against cardiovascular disease as a safe, cheap, and widely accessible adjunct to traditional medical therapies. Their review of sixty-nine clinical studies expressed concern that the quality of many studies was limited because the subjects were not appropriately randomized, lacked sufficient study population sizes, or had insufficient follow-up. The authors cautiously concluded that the evidence hints at a possible cardiovascular benefit, while acknowledging that the overall quality and, in some cases, quantity of data are modest. However, given the low costs and low risks of meditation, the study suggested that anyone interested may want to consider it "with the understanding that the benefits of such intervention remain to be better established." In other words, it won't hurt; it may help, but we don't know for certain if it is worth your time, cardiovascularly speaking.

The evidence is more promising regarding meditation's potential to reduce inflammation, whether through traditional practices or mindfulness-based interventions (MBIs). A study that reviewed forty-eight randomized controlled trials encompassing a total of 4,683 subjects found that compared to control groups, individuals practicing MBIs had decreased levels of C-reactive protein and interleukin-6, serum proteins that correspond to the amount of inflammation in the body. This study also found evidence for positive changes in telomere length, the molecules my Menla stage-mate Elizabeth Blackburn discovered that prevent your chromosomes from deteriorating and are related to the risk of inflammation, cardiovascular disease, and other conditions. It concluded that meditation may activate the inflammatory reflex, reduce inflammation, and protect your telomeres. While these findings are encouraging, they are not definitive proof of a causal relationship between meditation and vagus nerve activation.

Stimulating the vagus nerve directly with an electrode does indeed inhibit inflammation, and meditation has been shown to reduce inflammatory markers and inflammatory gene expression, but this correlation does not prove causation. We don't know that meditation stimulates the vagus nerve fibers, which are responsible for reducing inflammation. No one can say for sure that meditation causes the brain to activate the inflammatory reflex or to influence inflammation through some other pathway(s). We do not have the ability to test this idea in animal models because we cannot teach mice or rats to meditate. In human meditators, we do not have the ability to directly measure the electrical activity of the vagus nerve inflammatory reflex that suppresses cytokine production and inflammation. Absent direct experimental proof in animals and people, we cannot directly prove a cause-and-effect relationship between meditation and the inflammatory reflex.

This reasoning also applies to the potential role of the vagus nerve in the psychological benefits of meditation. I find it interesting that functional imaging studies have shown that many of the brain regions modulated during meditation are also modulated in patients with implanted vagus nerve stimulators. And that direct vagus nerve stimulation using an implanted device improves focus, cognition, and a sense of well-being in many, if not all, patients—benefits that many people say they derive from meditation. A cause-and-effect mechanistic link between meditation and the vagus nerve is certainly plausible, but as of yet, we lack direct proof. The final scientific word requires more study.

I was intrigued by the science suggesting meditation influences the brain, and I wanted to know what it feels like. So, nearly ten years ago, after watching a fascinating TED Talk by meditation teacher and author Andy Puddicombe, I downloaded an app and began meditating daily. I've meditated almost every day since, either guided by an app or on my own. There are many meditation apps you can choose from, but my personal favorites are Headspace (Puddicombe's) and Waking Up (from Sam Harris, philosopher, neuroscientist, and bestselling author).

While the scientific community continues to explore the mechanisms of meditation (and perhaps answer my questions about what is happening inside my brain), I've personally experienced its benefits. Most days, first thing in the morning, I practice a "focused attention" meditation with my eyes closed as I pay attention to my breath. For ten minutes or so, I note the thoughts, sensations, and emotions that come and go like monkeys in the trees jumping from branch to branch. Some days I practice a "loving-kindness" meditation, visualizing people I love and wishing them to be happy and free from stress and suffering. And other days I practice a bird's-eye meditation, envisioning myself first from the ceiling, then ascending higher and higher above the earth's surface, akin to taking flight, until I am outside of the solar system looking in. Everyday concerns and anxieties begin to look very different from outer space.

People are drawn to meditation for different reasons, ranging from stress reduction to improving mental health, enhancing focus, elevating athletic performance, and as a spiritual journey to foster inner growth. For me, I started meditating to experiment on myself.

I have learned a lot from my self-experiment, mostly that my preconceived notions about meditation, which were heavily biased by interpretations I'd come across over the years in print, popular media, and online, were wrong. I have not had and do not seek to have a religious experience. I have not found nirvana. I do not expect to "receive total consciousness," as was promised to Bill Murray's character in *Caddyshack*. And I do not expect to "empty my mind," a common misconception of "successful" meditation. But every time I do it, I am fascinated by just watching what happens in my mind for ten minutes at a time. I witness sensations, emotions, and thoughts appearing and disappearing, one at a time, each in their own present, and practice not reacting to them. I am endlessly intrigued by observing the thoughts and feelings that arise seemingly out of nowhere. I wonder who or what is doing the "paying attention." And I enjoy being simply present in the moment, without going backward to rehash some prior event or skipping forward to some feared or desired future.

After doing all these "experiments" on myself, I still have the same question. How does my brain decide what to pay attention to?

I have learned that it is not possible for me to decide what will appear in my mind's eye. I am more like a subway rider watching who gets on and off at each stop, or a bystander coming across a crime scene, waiting around to see what will happen next. When I do try to peer around the curtain to see who or what is deciding, all I find is open space. Curiouser and curiouser, then the bell rings, the ten minutes are up, and I am thrust back into my decision-making mode, back in charge of planning my next and every move for the upcoming twenty-four hours.

While I do not fully understand the nature of consciousness and attention, I take some solace in knowing that this is one of the three greatest unsolved mysteries in all of science, meaning that nobody understands it. (The other two mysteries being the origin of the universe and the origin of life.) I am okay living with the open question.

Since these are my own experiences and not the results of large randomized clinical trials, I won't make specific recommendations about meditation. But I have made it a habit, like brushing my teeth, and just do it, for several reasons. First, I believe my lab's results for the past twenty years showing that reducing inflammation, and therefore the damage it can cause, is good for my blood vessels and organs, including my heart and brain. Second, the clinical studies I have read, some of which I touched on above, suggest meditation can, in many people, if not everyone, reduce inflammation, heart rate, and blood pressure. Third, clinical studies also suggest meditation can modulate the activity of brain centers and enhance a sense of well-being and happiness for many people.

Finally, I like to be happy, and meditation seems to be a good way to practice staying in the present moment, less reactive to events or outcomes I cannot control.

BREATHWORK

You may be familiar with *diaphragmatic breathing* from self-help books, websites, podcasts, and other resources. It is often promoted as a way to activate your vagus nerve and enhance your vagal tone.

There is some science behind it.

Diaphragmatic or simply deep breathing is the conscious, deliberate deepening of your breath, engaging the diaphragm, a large muscle located at the base of your lungs. When you breathe deeply from your belly, your diaphragm contracts and moves downward, creating more space in your chest for your lungs to expand. This lung expansion activates signals traveling up the sensory vagus nerve (input). When these arrive in your brain, they activate an outflow of parasympathetic responses returning to the body via other vagus motor fibers (output) to produce a coordinated behavioral response including a slower heart rate, lower blood pressure, and a general sense of calm and enhanced well-being.

This body-mind integration method has been studied extensively as a nonpharmacological intervention for making people feel better, reducing anxiety, depression, and stress. Clinical research suggests it can be a useful treatment in some patients with PTSD, phobias, and other stress-related emotional disorders. It may even enhance the quality of life in healthy subjects. One study from Harvard Medical School in Boston and Beijing Normal University in China showed that healthy volunteers who underwent an eight-week training program of diaphragmatic breathing had improvements in cognition, emotional outlook, and vagal tone, as well as decreases in respiration frequency and negative affect scores and increases in sustained attention scores. In short, diaphragmatic breathing can promote your ability to relax and be happier, a finding supported by numerous studies across diverse populations.

The Wim Hof Breathing Method

The extreme athlete Wim Hof, or "The Iceman," as he is known for the world records he has set in extremely cold environments, is renowned for his breakthrough breathing techniques. Wim holds the records for time submerged in an ice-filled bath (nearly two hours), for climbing up to a height of 24,300 feet on Mount Everest wearing only short pants and shoes, and for swimming under ice for 188.6 feet. (He managed that last feat the day after a practice swim when his corneas froze, temporarily blinding him until they thawed.) I was fortunate to meet Wim, and even study him, when he visited our lab many years ago during a visit to the United States from his native Netherlands.

Exposure to cold is one pillar of Wim Hof's method, and we'll explore the relationship between cold and the vagus nerve in the next chapter. But he also developed a unique breathing method as a teenager during an almost spiritual quest. Here's how he talks about those early years:

> *Like many, probably, I was in a soul search and I visited many countries, traditions, languages, esoteric disciplines . . . karate and kung fu and yoga, but also the dervish, the Sufi, Buddhism, all kinds of traditions and disciplines, but it could not really fulfill me, in the depth.*

Wim read hundreds of books as a young man before concluding, "The answer is not in the head." Where is the answer, you might ask? He says, "The answer is in the body and the brain together."

Today, in his mid-sixties, Wim boldly proclaims on his website the premise of his method: "YOU ARE STRONGER THAN YOU THINK YOU ARE." He likes to use "the full capacity of [his] physiology," and he likes to make a statement. Wim says he "told everybody," decades ago, that "the autonomic nervous system and the immune systems related to the autonomic nervous system can be influenced," and that people mocked him at the time, but he was convinced. He has long believed that "the autonomic nervous system will be no longer autonomic," meaning that people could learn to influence its functioning at will. So he set out to obtain scientific proof.

Wim had arrived in New York from Amsterdam on a tour of the U.S. to raise interest in the methods he uses, which he believes are the key "to be happy, strong, and healthy." Intrigued by our research on the vagus nerve and inflammation, he exclaimed in a thick, booming Dutch accent when we met, "I want you to study me for science and the world."

Surprised and unprepared for this request, Sangeeta asked, "When?"

"Right now!" Wim shouted, with a grin that made it hard not to like him.

In forty-eight hours, which is the "right now" for clearing scientific research regulations, we received emergency permission from our Institutional Review Board, the committee that oversees human subject research, allowing us to proceed with a clinical experiment testing whether Wim could voluntarily regulate his own immune system using his then mostly unknown breathwork methods, as he claimed. Today, the Wim Hof Method, as it is taught globally by hundreds of people he has trained, is practiced daily by millions. But when we first studied him back in 2007, relatively few people knew him or his practice.

Wim Hof breathwork is similar to an ancient practice called *Tummo* meditation or "inner fire" breathing, which combines breathing exercises and visualization to generate internal heat. (Tummo is a Tibetan goddess of heat and passion.) Yogis in harsh Himalayan conditions have been observed to increase their internal temperature, even to the extent of drying wet sheets wrapped on their naked bodies in the cold mountain air. In contrast to the relaxing meditative practice of sitting quietly and following your own breath, the Tummo technique teaches breathing patterns that exercise the respiratory muscles, which generates body heat, combined with other cognitive meditative practices focused on visualizing heat arising in the spine.

Wim's breathing method involves a cycle of thirty deep inhalations followed by thirty relaxed, unforced exhalations. (This is important: Breathing this way can lead to lightheadedness, dizziness, or even fainting. If you try it, make sure you're in a safe position, sitting or lying down, away from water, and not using equipment that could harm you or others if you were to lose consciousness.) At the end of the final exhalation, subjects then hold their breath for an amount of time determined by individual tolerance ("until you feel the need to breathe again"). Next comes a deep inhalation or "recovery breath," as Wim calls it, that you hold for ten to fifteen seconds, then let go, before beginning another thirty-breath round.

In the clinical research center at the Feinstein Institutes, Wim lay back on the exam table and donated several tubes of blood to our study. Then, leaving him to his own breaths and thoughts, Sangeeta and I turned out the lights, shut the door, and went for coffee. Two hours later, we returned and drew more blood from the Iceman before shaking his hand, saying goodbye, and wishing him well until we saw him again on his return trip the following week.

When the lab results came in, Sangeeta said, once again, "Kevin, you have to see this."

"What did you do last week after we shut the door and left you alone?" I asked Wim a week later.

"I just breathed like a motherfucker!" he barked, laughed, and grinned.

However he had breathed, he had lowered cytokine production in his blood by almost 75 percent. It seemed he had voluntarily inhibited the inflammatory activity of his immune system in a way that we, and nobody else, had ever seen before.

Then he asked, "Are you going to publish me?"

I told him I couldn't publish him, because he was a single case. We needed much more evidence to publish. He continued to press, asking what it would take to publish this discovery.

I said it would be necessary to repeat this kind of study in twenty to thirty people divided into two groups. One group should be trained in his breathwork to see if their inflammation is suppressed compared to the other, untrained, control group. Then you would have to teach the control group to do it and check whether their inflammatory response is correspondingly inhibited. I offered to do this experiment for him, adding that to do it properly would require about $1 million to cover the costs of designing the study, statistical evaluation, and blood analyses.

I added, "I hope you do this, because if you are right, your work could help a great many people in this world."

A few years later, I was pleased to read a paper in the *Proceedings of the National Academy of Sciences*, two years before we would publish in the same journal our clinical trial results of using vagus nerve stimulation to suppress inflammation and treat rheumatoid arthritis. The paper summarized the results of a Wim Hof Method study performed in the laboratory of Peter Pickkers at the Radboud University Medical Center in Nijmegen, the Netherlands. They had studied two groups of people, as I had recommended, one of them trained in Wim's breathwork. And it had worked.

In Pickkers's study, they infused bacterial endotoxin into the subjects' veins, using methods similar to the experiments we had performed on healthy subjects on shore leave in New York. As expected, the untrained control subjects experienced flu-like symptoms due to high levels of TNF and other cytokines triggered by endotoxin. But the subjects in the breathing intervention group were significantly better off, with less severe symptoms and lower levels of TNF and other cytokines. The study also found that Wim Hof breathing reduced acid levels in the blood (called *respiratory alkalosis*), low oxygen levels (*hypoxia*), and significantly elevated epinephrine levels. These epinephrine levels were many times higher than what occurs during a typical fight-or-flight response. Furthermore, levels of IL-10, which is a cytokine that inhibits inflammation, were also increased as compared to the controls. The study's lead author, echoing Wim Hof's vision, concluded that "through practicing techniques learned in a short-term training program, the sympathetic nervous system and immune system can indeed be voluntarily influenced."

While this idea of increased fight-or-flight suppressing inflammation may initially seem counterintuitive, given the evidence that the sympathetic nervous system stimulates inflammation (as happens in depressed subjects with chronically activated fight-or-flight responses in chapter 7), the conundrum is explained when you understand that the duration and amount of sympathetic stimulation is what matters. Epinephrine and norepinephrine released by the sympathetic nervous system interact with a variety of specific neurotransmitter receptors on the surface of immune cells to control the production of cytokines and other inflammatory molecules. Some of these receptors, called *alpha* and *beta*, do inhibit the production of cytokines, but other receptors do the opposite because they stimulate the production of cytokines. During chronically elevated, but relatively low levels of sympathetic neurotransmitters, inflammation and cytokines are increased, and this is what happens in anxiety, depression, and other chronic conditions. However, during a sudden, intense release of neurotransmitters, as occurs during intense exercise, Wim Hof breathing, or a massive fight-or-flight response to an eighty-foot-high dive off the Mostar Bridge, the opposite effect is observed: Inflammation is suppressed.

This happens because acute, extreme increases of sympathetic neurotransmitters inhibit the immune system production of cytokines and other inflammatory molecules. The phenomenon was first clearly observed in experiments where volunteer subjects received intravenous infusions of high-dose epinephrine (a neurotransmitter) and cortisol (a stress hormone), to simulate a major fight-or-flight response. Then they received endotoxin and the researchers waited for the subjects to develop the expected flu-like symptoms caused by the accumulation of cytokines in their blood. Surprisingly, the simulated fight-or-flight infusions significantly inhibited the release of cytokines, preventing the severe flu-like illness typically observed in subjects infused with endotoxin alone. The researchers went on to further study the underlying mechanisms of these responses using a whole-blood cytokine assay, like the one Sangeeta and I used to demonstrate the effects of vagus nerve stimulation inhibiting cytokines in healthy humans.

When they put norepinephrine into the blood in a test tube followed by the addition of endotoxin, the white cells in the blood produced significantly less TNF and other cytokines. The study concluded that the enhanced release of norepinephrine provides a negative feedback mechanism that inhibits the production of TNF and other cytokines. This means that very high fight-or-flight hormone levels, like what happened with the Wim Hof breathers, suppresses inflammation. Again, it's counterintuitive, given the common association of stress and inflammation, but a sudden high dose of norepinephrine or big-time fight-or-flight turns *off* inflammation; it is low-grade, prolonged elevation of norepinephrine, as in chronic stress, that stimulates inflammation. This finding does not negate the vagus nerve's role in suppressing inflammation during severe stress, however.

Recall the sheep on a treadmill in New Zealand in chapter 2, when researchers studied what the vagus nerve was doing during their workouts. Those results overturned longstanding dogma that the sympathetic and parasympathetic systems are mutually exclusive, for example that fight-or-flight responses during exercise silence rest-and-digest responses. Instead, the New Zealand study revealed that the vagus nerve was indeed *activated* during exercise, and this activation contributes to optimizing the sheep's physiological responses by increasing performance in the heart and lungs.

There is a good reason that the role of the vagus nerve during fight-or-flight has not been better appreciated before. This is because in humans it is relatively easy to measure epinephrine, norepinephrine, and cortisol levels as evidence of the magnitude of the fight-or-flight response, but the same cannot be said of measuring acetylcholine, the "vagus stuff" that is the principal neurotransmitter of the parasympathetic nervous system. The challenge lies in the fact that acetylcholine rapidly degrades in the bloodstream. That means we cannot measure its levels or correlate these to vagus nerve activity. And further, as we saw in the jogging sheep, when your heart rate is racing along under the surge of epinephrine and norepinephrine during a fight-or-flight response, it is nearly impossible to detect the influences of the vagus nerve unless you measure specific parameters, such as coronary artery blood flow, which is done by implanting invasive monitors into the heart. Obviously, while these kinds of experiments are feasible in sheep, they are ethically challenging in healthy human volunteers.

Slow and Cyclic Breathing

While Wim Hof breathing activates stress responses to suppress acute inflammation, other research has shown that slow deep breathing, which stimulates the vagus nerve, can also reduce inflammation. So, breathwork triggers fight-or-flight, which reduces inflammation, *and* breathwork triggers rest-and-digest, which reduces inflammation? Yes and yes, both are true.

As the Harvard-Beijing study and others have shown, deep,

diaphragmatic inhalations followed by prolonged, slow exhalations decrease your heart rate and increase your vagal tone, a kind of internal vagus nerve stimulator, no device required. And at Stanford University, researchers studied three different five-minute daily breathwork exercises compared to an equivalent duration of mindfulness meditation over one month. They compared cyclic sighing, emphasizing extended exhalations; box breathing, involving equal durations of inhalation, breath retention, and exhalation; and cyclic hyperventilation with retention, featuring longer inhalations and shorter exhalations, somewhat akin to Wim Hof's breathing. Each of these breathwork exercises reduced the subjects' respiratory rate and improved their overall mood and sense of well-being significantly more than subjects who practiced meditation without breathwork.

The group practicing cyclic sighing had the most significant improvement in positive affect scores. The precise mechanisms remain uncertain, but the authors speculate that vagus nerve pathways to the brain, by carrying signals into the brain centers controlling emotion, may be the reason daily five-minute sessions of cyclic sighing help alleviate stress. Although the argument for cause-and-effect remains open, cyclical sighing with extended exhalations can transiently decrease heart rate and increase heart rate variability and decrease the production of inflammatory cytokines, which we discussed in chapter 6 as a possible cause or contributor to anxiety and depression.

Many dozens of other clinical studies have addressed the effects of slow deep breathing on inflammation. In one of these, thirty-six hypertension patients were trained using a cellphone app that guided them in slow breathing at a rhythm of six cycles per minute, five times per day for three months. This protocol resulted in a significant decrease in blood pressure and TNF levels (even though the training did not produce a lasting or measurable increase in vagal tone as assessed by HRV).

Slow deep breathing may even inhibit levels of cytokines during inflammation caused by COVID-19. A recent randomized controlled clinical trial studied forty-six patients hospitalized by severe respiratory illness from coronavirus 2 (SARS-CoV-2) infection. Patients were randomized to control and intervention groups, and the intervention group was instructed by a cellphone app to perform twenty-minute episodes of six breaths per minute, three times daily. Within two weeks, patients in the intervention group performing slow-paced breathing reduced one serum cytokine (IL-6) as compared to the controls, leading the authors to conclude that larger studies are necessary and warranted to replicate these results, because "slow-paced breathing could be an easy to implement, low-cost, safe and feasible adjuvant therapeutic approach to reduce circulating IL-6 in moderate COVID-19 pneumonia." To me, sixty minutes of breathing exercises sounds like a chore, but confinement to a hospital bed would be a powerful incentive to invest the time.

What else is there to know about the effects of breathing on the activity of the sympathetic and parasympathetic nervous systems, and these links to inflammation and emotional well-being? A lot. And we are only beginning to unravel the mysteries of the great nerve and its reflexes, because it's complicated.

Out of the one hundred thousand vagus nerve fibers on each side of your neck, perhaps a few hundred are required to control breathing, and a few thousand or fewer are required to control inflammation. Yet we don't know if these are the same or different fibers, and we don't know how they contribute to other protective, homeostatic, and healing reflexes and brain centers. The daunting complexity of neural signals traveling in these other more than 195,000 vagus nerve fibers represents a major challenge to unravel.

Let's do a thought experiment to illustrate this complexity. Perhaps, if we slow the breath to 6 breaths per minute, we are specifically engaging maybe 2,000 vagus fibers (a reasoned guess, which scientists call a *hypothesis*), and maybe each of those fibers (another hypothesis) directs signals to synapses in certain brain centers that activate 35 other fibers, 12 of which send signals to another part of the brain, and 23 send projections down to the spleen to inhibit cytokine production . . . and so on, with each next step requiring further investigation. From the brain, there may emerge signals traveling in thousands of other vagus nerve fibers, and other nerve fibers, that connect to the heart, lungs, pancreas, and immune system. But that is not all, because signals traveling in other nerves go the opposite way, traveling upward into the regions of the brain that influence emotion and a state of positive mindfulness. These hypotheses can all be tested, but because the experiments have not yet been conducted, what I know for sure right now is that there is still a tremendous amount of work left to do before we'll have scientific certainty about all these connections and their functions in inflammation and emotional well-being.

Meanwhile, as with meditation, I await answers that may or may not come in my lifetime. And again, because I believe my lab's data, and have confidence in the findings from the clinical studies I have reviewed, and because there isn't any downside, I do incorporate breathwork into my routine. Most days I do several minutes of cyclical breathing, beginning with a large inhalation, followed by a short further inhale through my nose, like a big sniff. I follow this with a long, slow exhalation through pursed lips, producing an inhalation phase of about three seconds and an exhalation phase of seven seconds, meaning I am breathing at a rate of six breaths per minute. My four daughters are all grown up now, but if you have younger children and are pressed for quiet time alone, they can do this kind of breathing exercise with you. Breathwork is taught in some schools, based on data that indicates it is feasible to implement and is well tolerated by the students.

A few times a month, usually near the end of a workout session, I practice Wim Hof breathing. I never do this in a pool or near water or while I'm driving, and neither should anyone else, because there's a risk of fainting. I lie on the safety of my gym mat and take 30 deep inhalations to fill my lungs. Each inhale is followed by a fairly rapid exhale, emptying my lungs through my mouth or nose in what I will call a relaxed gust, and allowing my diaphragm within my belly to expand and contract naturally with each breathing cycle (Wim recommends wearing loose clothing). After the 30th exhale, with my lungs empty, I hold my breath for as long as I can. Usually, after the first 30 breaths, I can hold my breath for about 90 seconds. When the need to breathe feels overpowering, I inhale and continue to hold my breath for 10 seconds more. After exhaling, I start another round of 30 breaths and repeat the process. With practice, after several rounds, I have found it is possible to hold my breath for three or even four minutes. According to Wim, "It feels wonderful."

Not sure I completely agree, but after three rounds, on those days when I can hold my breath for three minutes or even longer without panic or discomfort, I feel like I am watching someone else do it, and that's an interesting feeling. And afterward, when I am "fully charged," as Wim says, I find I can do more than my usual number of push-ups, which makes me happy too.

"We are not making use of the full capacity of our physiology," Wim laments. "Now we found out we've got different layers," he says, "and we never use [our full capacity]. And this is the way to use it, to tap in and bang. Into the primitive brain, into the endocrine systems, immune systems, the way nature has meant it to be. Everybody is able to do it."

Wim's enthusiasm is real, and there is some good science behind it. But we still have a great deal more to learn about these methods. So, to be safe, just as you should do before engaging in any new regimen of strenuous exercise or overexertion, first check with your physician.

I am a single case study, so the same results may not apply to you. But I do enjoy how meditation and breathwork make me feel, and I suspect these practices are limiting the effects of inflammatory damage to my body and blood vessels. Lacking incontrovertible proof, I have adopted a Pascal's-Wager strategy when it comes to both meditation and breathwork: In case they are good for me, and since I don't mind doing them most days, then I may be blessed with better health along the way. But if I don't do these things, and become stricken with illness I could have avoided, I will regret not having done them.

10

Cold and Exercise

The cold is a teacher. It's merciless. You don't picnic when you go into the cold. You don't think about your mortgage or your kid's braces or your divorce; you just survive. You reactivate the deepest part of your brain.

— WIM HOF

On an unseasonably warm spring day in Old Lyme, Connecticut, when we were teenagers, my brother Tim and I finished our chores and decided to go for a swim. After working several hours in the sun, washing windows and raking the lawn at the family beach house, we decided a quick dip was necessary to cool off. Raising his eyebrows in wonder and disbelief, my father just nodded, knowing the water temperature was glacial. But he did not protest as we put on our swimsuits and headed to the shore. Tim arrived first and was already standing in the waves. "How's the water?" I shouted, running toward him across the sand.

"Warm! Great!" he lied.

With a running dive, I flew past him. But then, in midair, I noticed Tim breaking into a sprint headed back to the house.

What happened next unfolded in slow motion, like a car crash when every second is a freeze-frame. My thoughts, my emotions, and I were suspended together, hanging in the air over the 42 degrees Fahrenheit waves.

My head hit the water first, followed immediately by pain in every square millimeter of my body. After fighting my way back to the surface and standing waist-deep, I felt my entire body exhale all at once, emitting a pitiful sound as the final molecule of air exited my lungs. I never knew it was possible to exhale so quickly, involuntarily, or completely. I wanted to inhale, but I couldn't. The trickster Tim, who had only walked into the water up to his ankles, was halfway home, laughing until he cried.

For what felt like an eternity, I could not will my lungs to life. When I could finally breathe, I bolted after my brother, intent on wrestling him to the ground and exacting revenge. Even now, I smile every time I think of it.

That first cold plunge happened decades before I would adopt the practice into my own daily routine, and long before a flood of social media posts tagged with #coldplunge would rack up billions of views. In part because of the expanding popularity of Wim Hof, the Iceman, whom we met in the previous chapter, and many others who also extoll the health benefits of cold exposure. Hollywood stars, podcast hosts, professional athletes, social media influencers, wellness practitioners, and life coaches provide a steady drumbeat promising benefits of the cold to hundreds of millions of followers on media outlets and in books and other publications.

These practices, called *cold plunging, cold water immersion,* or *cold therapy,* are supposed to reduce inflammation; alleviate pain in the joints and muscles; improve oxygenation; enhance the removal of metabolic waste in tissues; promote muscle repair; stimulate metabolism in brown adipose tissue (brown fat) to burn calories and generate heat; increase metabolism to aid in weight loss; reduce insulin resistance; normalize lipid and glucose levels; increase mental resilience; reduce stress; improve mood; enhance alertness, euphoria, and relaxation; rejuvenate skin . . . It's quite a long list and it doesn't stop there, but I will, due to page limitations.

The key takeaway is that most of these touted benefits of cold exposure are based on study results that are closer to anecdotes than science. The most popular claims remain largely unsupported by rigorous data from randomized, well-controlled clinical studies. Absent this evidence, coupled with the all-too-real dangers of cold exposure in people with underlying cardiovascular and pulmonary conditions, healthcare professionals don't tend to include cold exposure as a standard part of evidence-based wellness regimens. Thus, we are confronted by a significant disconnect between the huge number of anecdotes purporting benefits of the cold and the absence of broader medical recommendations. Gold standard evidence could bridge this divide.

To evaluate the effectiveness of any treatment or intervention, we need randomized controlled trials (RCTs) designed to minimize bias and confounding variables by randomly assigning participants to different groups. Typically, one group (the intervention group) receives the treatment being studied, while the other group (the control group) receives either a placebo or standard treatment. For example, in Ulf Andersson's study of transcutaneous auricular vagus nerve stimulation (taVNS) both groups will use a device on their ear, but only one group's device will administer electrical stimulation. The other group doesn't know it, but their "treatment" is a placebo. By comparing outcomes between the two groups, researchers can determine whether the treatment has a statistically significant effect. For this to occur, there must be enough subjects in each group to enable a meaningful analysis. Large-scale RCTs are the gold standard, but smaller clinical trials can also play a valuable role in developing new ideas in the research process.

Clinical trials enrolling small numbers of people can be valuable for generating initial observations, new ideas, and hypotheses. But they are often viewed with skepticism and interpreted as "preliminary" because of the small sample sizes, variability in participant characteristics, and potential biases. If you and I (n = 2) each cold plunge once a day for three weeks, we can't be sure if any changes we experience are due to cold plunging itself or to other factors that make us similar or different. That is why the results from smaller clinical trials of cold exposure therapies, even when they include comparisons between intervention and nonintervention groups, are considered preliminary and are referred to as "anecdotal evidence." A clinical trial of one or two subjects is more like a fairy tale with a happy ending than scientific proof that the cold makes you healthier and stronger. Personal experiences are not enough to predict with any certainty how cold plunging will affect anyone else. The more an intervention or practice is observed to produce reproducible effects in large numbers of diverse people in a randomized controlled trial, the more confidence we can have that the same practice will be effective when other people do it.

After the first large, well-designed RCTs of cold therapy pass muster with sufficient statistical power (the ability of a study to detect a real effect if one exists), the next step in medical decision-making calls for additional, independent RCTs to replicate the results and confirm the robustness of the findings. Until this happens, and this may take several years, physicians will remain loath to generalize recommendations about using cold therapy to broader populations.

I believe Wim Hof when he says cold exposure made him healthier and happier, but whether that means it will do the same for you and everyone else remains an open question. Current advocates of cold practices may cry foul, but this conservative approach to medical decision-making is grounded in Hippocrates's oath to first do no harm. As it should be. The replication of large RCTs is the best method we have to balance the benefits of cold interventions against potential risks and to accurately predict the outcomes in the broader population.

As a scientist, I am intrigued and inspired to study further the findings of smaller clinical trials of cold therapies that do provide preliminary evidence for their benefits. And I encourage my colleagues to think about launching larger RCTs to build evidence for the risks and benefits of cold plunges. Demand is high for more RCT data on cold therapy, evidenced by the website clinicaltrials.gov, which lists hundreds of trials studying ice baths and other modes of cold exposure. Together with the medical and scientific communities, I await the results of these studies in the coming years. So with these caveats and disclaimers in place, let's have a closer look at some of the results from lab experiments and smaller clinical trials that I find so interesting. And let me explain why I personally take cold showers.

COLD PLUNGES

Galen, the first physician-scientist and early explorer of the great nerve, was among the first to wonder about the effects of cold exposure on the body's physiology. He treated malaria patients by immersing them in cold water, a practice predating the efforts of countless physician-scientists who even now continue to study the uses of cold therapy.

More recently, we learned that cold stimulates specialized sensory neurons in your skin called *nociceptors*, which carry proteins on their cell surface called *transient receptor potential* (TRP) channels. Nociceptors work like the thermometer in your home's thermostat because when they sense cooling in your environment, these neurons send signals to your brain calling for heat. In severe cold, they send pain signals, which explains the extreme discomfort I felt when Tim tricked me into my first cold plunge. Like other sensory inputs, these nociceptors form the sensory arcs of protective and healing reflexes that protect your body from injury in cold environments.

The arrival of the incoming cold signals in your brain stem triggers reflexive, outgoing signals. Severe cold culminates in extreme fight-or-flight responses. If you fall through the ice and are suddenly immersed in freezing water, these sympathetic responses give you the surge of energy you will need to pull yourself out, powered by high levels of epinephrine and norepinephrine released into your bloodstream. These stress hormones also elevate your blood pressure to pump more oxygen to your muscles and produce vasoconstriction, a narrowing of your arteries that reduces blood flow to your skin to help keep you warm by diverting blood into your internal organs. This helps warm up your insides by reducing heat loss through your skin.

You might also expect this sympathetic response to increase your heart rate, and at first, it does. However, after several minutes remaining in the cold, the opposite happens. Your heart rate will slow. In one study in the early 2000s, ten healthy volunteers in Finland, clad only in shorts, socks, and sneakers, spent two hours a day for ten days in a 10 degrees Celsius (50 degrees Fahrenheit) refrigerated room while researchers measured their autonomic nervous systems. By the end of the two-hour chilling, when their skin temperatures had dropped to 88 degrees Fahrenheit, the subjects had signs of heightened fight-or-flight responses, evidenced by increases in blood pressure and norepinephrine levels, which is what the scientists had expected. But what they did not expect to see was that the cold had also activated the subjects' parasympathetic nervous system, evidenced by decreased heart rates and increased HRV. They discovered that the cold stimulated the subjects' vagus nerves—part of the protective reflex response known as the diving reflex.

In 2008, scientists first observed a combined response of the sympathetic and parasympathetic nervous systems in humans during whole-body cooling. (This finding predated later studies on the vagus nerves of sheep treadmill jogging in New Zealand, which also revealed sympathetic and parasympathetic cooperation.) The lead researcher noted that "the observation that both sympathetic and parasympathetic activity is increased during whole-body cooling is interesting, as this has not been documented before." The data indicated that in time, as your body habituates to cold, sympathetic activity declines and parasympathetic activity increases. As vagal tone increases, heart rate slows.

During extreme cold, a nuanced harmony between the sympathetic and parasympathetic nervous systems optimizes the delivery of warm blood and oxygen to your organs. Again, the sympathetic system's fight-or-flight effects are easier to observe and measure because your pulse races. And if you happen to be experiencing this in a lab where you can donate your blood, epinephrine and norepinephrine can be easily measured with specific lab tests. They leave behind lasting traces, whereas acetylcholine, the major neurotransmitter of the parasympathetic vagus nerve, rapidly dissipates, leaving few traces of its involvement in the body's response to cold. Vagus nerve cardiac influences, which tend to slow the heart, are also impossible to observe in this situation since they are masked by the tachycardia from sympathetic overdrive. But we now know that during extreme cold, your vagus is also working hard, in harmony with your sympathetic nervous system.

Your vagus nerve is also activated by breathing. And breathing is another physiological system influenced by cold exposure, something I learned that day as an unsuspecting teenager at the beach. Coming up to the surface of Long Island Sound, I experienced an episode of massive hyperventilation when the air in my lungs was involuntarily, suddenly, and painfully expelled. This was followed, for another minute or more, by my feeling unable to inhale, a state called *hypo*ventilation. While the initial shock of cold water can cause involuntary hyperventilation, deliberate breathing techniques can help to regulate the body's response to cold exposure.

Not everyone experiences these same respiratory responses to extreme cold. Some cold-exposed subjects do not spontaneously develop hyperventilation, but studies of volunteers who deliberately hyperventilate reveal that it increases heat production in the muscles that do the work of breathing, the intercostal and pectoral muscles. The heat generated by voluntarily hyperventilating can help maintain internal body temperatures and provides your tissues with more oxygen, the critical fuel your cells need to do work and generate heat.

We know that engaging in deep breathing exercises, such as diaphragmatic breathing, enhances heart rate variability (HRV) by activating the diaphragm muscle and stimulating the vagus nerve. One paced breathing method to increase HRV without hyperventilation is to inhale slowly through the nose for three seconds, followed by a slow, seven-second exhalation through pursed lips. Such deliberate breathwork reduces heart rate and increases HRV through the vagus nerve reflex called *respiratory sinus arrhythmia* (RSA) because during inhalation, the heart rate tends to increase slightly, and during exhalation, it decreases from increased vagus nerve signals that slow the heart. Many cold immersion practitioners use this three/seven breathing method to "stimulate their vagus nerve," which helps them habituate to longer periods in low-temperature environments.

Wim Hof says "the magic began" when he changed his breathing during exposure to cold. I do not doubt this, for him, but we need more studies to understand how it will affect you, and all of us.

COLD AND INFLAMMATION

You twist your ankle, running in stupid shoes to catch your train, so you put a bag of frozen peas on it when you get home. Your child takes a bad, fast pitch in the shin at baseball practice, and you keep some Pokémon cold packs in the freezer for just such an occasion. Cold therapies for local tissue injury using ice packs and other cooling methods are a staple of physical therapy and home care to reduce pain, swelling, and inflammation. They work because cold-activated fight-or-flight stress responses stimulate blood vessel constriction, which reduces swelling in acute injuries or inflammatory conditions.

One French study enlisted thirty-two patients with arthritis who volunteered to have fluid withdrawn from their knee for analysis. The researchers reported that cooling the swollen knee not only reduced the severity of joint pain but also significantly reduced the amount of pro-inflammatory cytokines and other molecules in the knee fluid. The local tissue benefits continue after the cooling stops because deliberately or spontaneously rewarming the cooled tissue causes the blood vessels to vasodilate, meaning their diameters widen, which increases blood flow to the affected area. This improved circulation helps promote tissue healing by increasing the availability of white blood cells, stem cells, and other factors that aid in recovery and restoring normal function. The frozen peas are a reasonably safe, low-tech way to obtain some relief from the pain and inflammation of these kinds of injuries and conditions.

It can be tempting to follow the logic that if a little is good, then more is better, but science would ask, a little of what? And science would test the hypothesis to find out if and how much more actually is better. If some cold applied on your knee stops inflammation there, will large amounts of cold on your entire body stop inflammation everywhere? Because billions of web impressions have lately been saying yes, I want to consider whether evidence proves the hypothesis that whole-body cold exposure reduces inflammation and inflammatory molecules, and whether cold-activated vagus nerve stimulation can reduce inflammation in your body.

Activation of your sympathetic nervous system, as occurs during cold stress, hyperventilation, and even vigorous exercise, can either inhibit or stimulate inflammation, depending on the timing of the response and its magnitude. Add to this complex picture the coactivation of the vagus nerve, which inhibits inflammation. So, what happens in the cold? Does the sympathetic or parasympathetic effect predominate? And does the outcome of this cooperation of autonomic systems stimulate or inhibit inflammatory molecules like cytokines?

Studies of various types of cold exposure, ranging from complete immersion in freezing cold water to standing in a refrigerated room, report a wide range of immune system cytokine responses. Consider a study somewhat paradoxically published in a journal entitled the *International Journal of Hyperthermia*. It involved twelve subjects who agreed to immerse themselves in very cold water (14 degrees Celsius, or 57 degrees Fahrenheit) for ten minutes. Within seconds after entering the water, the subjects' heart rates increased by as much as thirty beats per minute (bpm), a response that coincided with a massive surge in blood levels of the stress hormones epinephrine and norepinephrine, along with lesser increases in cortisol. Electrical recordings of the subjects' pectoralis major muscles indicated a sharp increase in muscle contractions, meaning these muscles were performing significantly more work to support breathing and hyperventilating, the same muscles that forced all the air out of my lungs when I took my first cold plunge. Despite these significant physiological responses to cold, when the investigators measured cytokines in the blood, they found only minor changes, some of which were slightly increased and others slightly decreased.

In another study from Prague, ten subjects volunteered to be submerged up to their necks for one hour in 14 degrees Celsius water (57 degrees Fahrenheit) *three times a week for six weeks*, an experimental design that aligns quite closely with many protocols advocated on social media for at-home chill-yourselfers. This study found, again, that increased muscle activity from the greater effort of breathing, combined with significant fight-or-flight responses with several hundred percent increases in stress hormones, produced only minimal or minor changes in the levels of cytokines, including TNF, which tended to increase, not decrease, slightly. The authors concluded that cold water submersion is a minor stress (which seems to grossly underestimate the stress I felt). They stated that "repeated cold water immersions, which increased metabolic rate due to shivering *and* the elevated blood concentrations of catecholamines activated the immune system to a slight extent." The biological significance of the changes they observed, they said, "remains to be elucidated."

Disappointingly, the reported changes in cytokine levels in these two studies are quite small and fall within the normal range of cytokine levels observed in healthy subjects. This is a very different scenario from the experiment Sangeeta and I performed when we were studying the effects on inflammation of vagus nerve stimulation in volunteers who underwent surgery to have a device implanted as treatment of epilepsy. In our studies, you'll recall, we collected tubes of blood containing white blood cells before and after the vagus nerve stimulator was turned on. Then we activated the white blood cells in the tubes by adding bacterial endotoxin, waited several hours for the white blood cells to manufacture new cytokines, and finally measured the freshly produced cytokines in the tubes. Our study design enabled us to address whether the white blood cells made more or less cytokines after receiving signals from the stimulated vagus nerve. Because we observed significantly lower cytokine levels after the vagus nerve stimulator was turned on, we concluded that vagus nerve stimulation inhibits cytokine production in humans.

But we do not yet have comparable results from subjects exposed to extreme cold. Simply measuring background levels of cytokines before and after cold exposure cannot answer questions about whether white blood cell functions are suppressed. We have relied on studying white blood cells in whole blood to bridge this gap because, with this approach, we can measure the ability of these cells to make new cytokines.

Another study that has taken this approach provides insights into the cytokine-producing activities of white blood cells in cold-exposed humans. In it, nine men dressed in light clothing were exposed to wind, rain, and cold (5 degrees Celsius, or 41 degrees Fahrenheit; 20 kmh, or 12.5 mph wind). As in the study we did in my lab, the researchers stimulated whole blood with LPS and measured the production of cytokines produced by the white blood cells using flow cytometry, a powerful analytical technique to quantify the amount of cytokines made in each individual white blood cell. This small study found that exposure to extreme cold (and wind and rain) significantly impaired the white blood cells' ability to make TNF and other cytokines. This impairment could be due to the cold stimulating the subjects' vagus nerve inflammatory reflex, which in turn inhibited the production of cytokines in their white blood cells, as my lab observed in the study of arthritis patients treated with a vagus nerve stimulator.

Here's the summary of what we can be confident that we know so far: First, whole-body cold is a major stressor, which can be dangerous to some people with underlying cardiovascular conditions. So, you should ask your doctor before exposing yourself to extreme cold.

Second, the initial response to extreme cold activates your sympathetic nervous system's fight-or-flight response and co-activates your vagus nerve. Shivering and vigorous effort from hyperventilation generate muscle work, which can stimulate inflammation. The surge of stress hormones can also both stimulate and inhibit inflammatory cytokine production, as we saw in the Wim Hof breathing trained subjects in the last chapter.

Third, if cold exposure persists, your vagus nerve will be stimulated, gradually slowing down your heart. This vagus nerve stimulation may also significantly inhibit inflammatory cytokine production. But this final point remains conjecture. It's an interesting hypothesis requiring further study from randomized controlled trials, which I certainly hope someone somewhere pursues carefully.

Until then, despite what you see on social media, the effects of cold exposure on the immune system and the vagus nerve's role remain unclear. We also lack sufficient data on the long-term benefits or harm from repeated cold exposure. So be careful what you believe from online hullabaloo and ask your doctors what they think.

Meanwhile, as with meditation and breathwork, while I await the evidence from additional studies, I take cold showers. Why?

Because I find some confidence in the consistent conclusions of the clinical studies that voluntary activation of a fight-or-flight response by cold exposure inhibits inflammation. And inhibiting inflammation is a good way to reduce the risks of cardiovascular diseases, neurodegeneration (including Alzheimer's), obesity, diabetes, and cancer.

So, three or four days a week, I end my regular shower by turning the water to full cold for two or three minutes. Once again, I cannot make recommendations about what you or anyone else should do based on my experience. But I have noticed a few interesting things about this practice.

First, it is very uncomfortable. Every single time. After hundreds of episodes over several years, I continue to feel stinging pain from the cold water (about 11 degrees Celsius, or 52 degrees Fahrenheit). But strangely, as the cold water beats down, after thirty seconds or so, I notice fewer negative emotional reactions to it. It seems as if I can witness the pain signals arising from my skin's nociceptors and arriving inside my brain, but at the same time, not react out of indignation, rage, or panic. It's like I'm observing someone else endure the cold. This mindset, when emotions do not have their usual sway over my consciousness and pain is not "unpleasant," is freeing.

Second, when it's over, I smile, every time. And as I have said, I like to be happy. I am not sure why this occurs. Perhaps the vagus nerve stimulation is sending signals up into my brain that contribute to a positive outlook. Or maybe daily vagus nerve stimulation helps lower the total bodily burden of inflammatory cytokines, reducing their negative consequences on my affective brain. Or maybe the afterglow of fight-or-flight combined with cold-mediated vagus nerve stimulation feels good. Or perhaps it's the relief of surviving the cold. Science hasn't decided, but I'll keep doing this thing that makes me smile.

Finally, I have noticed one other thing I think is interesting: I'm more comfortable outside when the New England winter weather turns cold. Since beginning my cold showers, winter cold does not bother me as much. Perhaps my regular cold-water practice has enriched my brown adipose tissue (BAT), the unique fat cells rich in iron that generate heat. Brown adipose tissue is activated by cold, converting glucose and fat into heat, potentially explaining how cold lowers blood glucose and increases insulin sensitivity. It is possible that long-term cold exposure can increase BAT, which can increase the energy expenditure of your whole body. A major factor controlling the activity of BAT is norepinephrine, produced during fight-or-flight responses to the cold, exercise, or the appearance of a lion in your living room.

One possible bonus: Norepinephrine is a major activator of brown adipose tissue, which raises whole-body energy expenditure. And this can help with weight loss. Although questions about cause and effect remain, normal-weight people do tend to have higher levels of BAT activity as compared to obese subjects. Whether this BAT activity explains the lower weight or whether the onset of obesity suppresses BAT activity is something scientists are still studying as they attempt to establish a causal relationship.

Perhaps cold showers are increasing my BAT, or perhaps not. Either way, tomorrow will be another cold shower day.

EXERCISING YOUR VAGUS NERVE

Exercise is good for you. We all know this. Robert Butler, the founding director of the National Institute on Aging (NIA), perhaps said it best when he said, "If exercise could be packed into a pill, it would be the single most widely prescribed, and beneficial, medicine in the nation."

Exercise has far-reaching and highly promoted benefits for mental and emotional wellness and for reducing the risk of cardiovascular diseases, obesity and diabetes, Alzheimer's and other neurodegenerative diseases, and cancer. When we consider that inflammation plays a causative or contributing role to all these conditions, which account for two-thirds of global deaths annually, and that exercise stimulates the vagus nerve, the link between exercise, the vagus nerve, and inflammation seems worth exploring.

We are inundated with advice to get off the couch, get up from our desks, and get out and exercise. Many of us feel we are not doing it often enough. What most people don't realize is that despite all this talk about the benefits of exercise, the exact mechanisms behind exercise's positive effects are not fully clear. If you're looking for motivation, you can find countless other resources available from health experts, body builders, coaches, athletes, exercise apps, and media influencers about how and when to exercise. In this book about your vagus nerve, which slows your heart rate and reduces inflammation, I want to get into some of the evidence linking exercise to increasing the activity of your vagus nerve and decreasing inflammatory responses and address whether this may explain some of exercise's benefits.

Regular, moderate-intensity, aerobic exercise—like brisk walking, elliptical training, cycling, rowing, or gardening, which increase your heart rate and breathing while allowing for sustained activity and conversation—has been linked to slower resting heart rates and higher heart rate variability (HRV) as compared to committed couch potatoes. Numerous studies, including an influential one at the University of Washington in Seattle, have established this link. They studied twenty-four men ranging in age from twenty-four to eighty-two. After six months of supervised bicycling, walking, and running, the subjects' heart rates were significantly decreased, and their HRV significantly increased. The investigators concluded that "exercise training increases parasympathetic tone at rest in both the healthy older and young men, which may contribute to the reduction in mortality associated with regular exercise." That conclusion, published in 1998, the same year my colleagues and I discovered that the vagus nerve's inflammatory reflex inhibits inflammation, is enough to motivate me to exercise more. I want to be in the slow heart rate group with those amped-up sixty- to eighty-two-year-old study subjects.

Numerous large studies confirm the benefit of exercise, like one study of 122,000 people over eight years showing reduced all-cause mortality in proportion to increasing cardiorespiratory fitness. The investigators in this study also found that the increased mortality risk from being out of shape, as assessed on a standardized exercise treadmill test, was comparable to the well-known risks of smoking, or having other conditions including diabetes and coronary artery disease. The exercise benefits extended into the later years of life because even the fittest seventy-year-olds outlived their less fit peers. However, while it is certain that regular aerobic exercise improves cardiovascular fitness, decreases resting heart rate, and improves vagal tone measured by HRV, proving that exercise is beneficial because it reduces inflammation is not so simple. There are many reasons for this, beginning with how we measure cytokines and other inflammatory molecules, as we have touched on. The problem is there are no gold standard blood tests of whole-body inflammation.

To understand what this means, consider the example of hemoglobin A1c, a blood test that you probably had done during your last annual physical examination. Hemoglobin A1c (popularized by Tony Cerami, my collaborator on monoclonal anti-TNF antibodies forty years ago) is a molecule that accumulates slowly in your bloodstream in proportion to the amount of glucose that also passed through your bloodstream in the preceding weeks. Hemoglobin A1c levels increase when the total amount of glucose passing through your bloodstream is high. This is critical for diagnosing diabetes since glucose levels spike up and down quickly, so it is possible to have a low blood glucose level at the time your blood is being tested while at the same time having a very high hemoglobin A1c level. Since the hemoglobin A1c level reflects all the high and low spikes in the preceding days and weeks, it is the gold standard for measuring your risk for diabetes.

But there is no "hemoglobin A1c similar" test for inflammation that we can use to look at the total burden of inflammation that your body has experienced in the prior days and weeks. Although we can indeed measure cytokines in your bloodstream, like glucose, these inflammatory molecules come and go rapid fire, sometimes being very high, but sometimes very low. Since cytokine highs and lows occur both when you are free from inflammation and when you are full of inflammation, the predictive value of a single cytokine measurement is very low to nonexistent.

That is why in our lab, Sangeeta and I stimulated whole blood with bacterial endotoxin to activate cytokine production by white blood cells. In this way we can assess not the variable background levels of cytokines in the blood, but rather the amount of cytokines that white blood cells are able to produce before and after vagus nerve stimulation. This distinction is important because after white blood cells receive signals from a stimulated vagus nerve, they make much less cytokines, and this impairment persists for many hours or even days. That is a more stable method to study the influence of the vagus nerve on cytokine production. Unfortunately, however, only a few smaller studies have employed these same methods to understand the effects of exercise on cytokine production, and there is significant disagreement among the results.

In one study of sixty-five eighty-year-olds, researchers at Purdue University compared a physically active group to a physically inactive group using flow cytometry to look at the amount of cytokines in individual white blood cells. After twelve weeks of exercise, they observed that the cytokine-producing monocytes were significantly decreased. They also found that endotoxin-stimulated whole blood produced less TNF, much as my lab observed during vagus nerve stimulation.

I collaborated on another study with Richard Sloan and other colleagues at Columbia University in New York, who enrolled sixty-one previously sedentary adults into moderate and higher intensity groups, defined by exercising at either 55–60 percent or 75–80 percent of their maximum heart rates, respectively, for a twelve-week period, then measuring endotoxin-stimulated whole blood TNF production. While the aerobic training program increased the fitness of both groups, as measured by VO2 maximum testing, white blood cells in the high-intensity group made less TNF, whereas the moderate-intensity group produced normal (higher) amounts of TNF. These findings suggest that the intensity and duration of your exercise regimen may determine whether exercise suppresses cytokine production by your white blood cells, or not.

An overwhelming amount of clinical evidence indicates that consistent, vigorous exercise enhances HRV and vagus nerve activity and is beneficial for the body. But regular exercise is also associated with improvements in brain and emotional health, as evidenced by studies linking exercise to improved cognitive function, reduced risk of anxiety, depression, and age-related cognitive decline, and enhanced longevity.

Why this happens is unclear. There are many possible explanations, some including a role for the vagus nerve. It could be that exercise increases the sensory vagus nerve signals that activate brain regions associated with enhanced positive emotions and a sense of enhanced well-being and happiness. Another possibility is that vagus nerve signals traveling away from the brain and back into the body decrease inflammation and inhibit the production of cytokines that cause depression, anxiety, fatigue, and negative emotions. Since exercise can suppress these depression-causing inflammatory factors, it stands to reason this could improve your outlook and enhance your sense of well-being. Or perhaps signals traveling from the body to the brain in the sensory vagus nerve somehow inhibit inflammation in the brain, which may also improve emotional outlook. Among other plausible explanations, exercise may be increasing the amount of brain endorphins, the body's natural narcotics that alleviate pain and induce feelings of pleasure or euphoria. But there is no consensus on these mechanisms.

More work is necessary to help us understand why exercise is beneficial and whether and how the vagus nerve plays into these benefits. Meanwhile, I try to exercise five days a week for thirty to forty-five minutes, combining aerobic exercise, stretches, and yoga with resistance and weight training. Strength training preserves muscle mass, improves metabolic function, and prevents injuries. Aerobic exercise, such as walking, cycling, or swimming, enhances cardiovascular health and endurance.

Some of the people who seem to be most trustworthy on this subject, like the Stanford University Medical School–educated and Johns Hopkins University–trained physician and bestselling longevity author Peter Attia, also advise us to listen to our body, adjusting exercise intensity as we need to and finding activities that are enjoyable and sustainable over the long term. Whoever first said the best exercise routine is the one you actually do may have also said it best.

Whether through meditation, breathwork, cold exposure, exercise, or all of the above, your vagus nerve will thank you for paying attention to it. I wish you the best on your physical and mental journey toward a healthful life. Long may you and your great nerve wander toward less inflammation and enhanced homeostasis and happiness.

11

Your Great Nerve

How to Talk to Your Doctor (the FAQs)

At one point I felt so hideously isolated that I called the doctor up and he said, "It's not the Dark Ages. There's a lot of options. Why don't you come in and we'll go through all the options?" That's what I did. I listened and I chose what was right for me.

— CYNDI LAUPER, ADVOCATING FOR PEOPLE
WITH PSORIASIS, AN AUTOIMMUNE DISEASE

Your vagus nerve is great because it reaches into so many life- and health-giving systems in your body and keeps them all in balance, ideally. Its complexity and range of activities are too much for one book to cover exhaustively—at least, one that isn't too big and heavy to lug around. And as in all scientific fields, vagus nerve research is work in progress. Your vagus nerve continues to harbor mysteries that, once solved, will open new possibilities for a healthier future. This is what most excites my colleagues and me in the lab, and it means that this book is a teaser of possibilities.

At the same time, bioelectronic medicine is a new enough frontier that if you're looking for a treatment for yourself or someone you love, you may face ignorance and doubt about proven therapies and reasonable applications, even (or especially) among medical professionals, who are used to the ways they've always done things.

Is there a vagus nerve therapy for what's bothering you? A vagus nerve treatment for your diagnosis? I cannot, of course, answer that without knowing you. But in this chapter, I'll walk you through some of the right questions to ask yourself and your doctors so that you can navigate the options and make the best possible informed decisions for your optimal health and happiness. These are not so much frequently asked questions as questions I hope more people will feel empowered to ask, along with answers you can use as a reference for yourself or someone you care about when having these kinds of conversations. They are, I hope, frequently asked questions to be.

What is vagus nerve stimulation (VNS)?

Vagus nerve stimulation (VNS) is an FDA-approved medical therapy delivered by a medical device implanted under your skin, usually by a neurosurgeon. The VNS device delivers electrical impulses to your vagus nerve, which runs from your brain through your neck and into your chest and abdomen. Because your vagus nerve lies under several inches of skin and muscle, surgery is required to connect the device's stimulating electrode directly onto the nerve.

How does it work?

The VNS technology is a battery-powered computerized device programmed to deliver regular small pulses of electrical current into your vagus nerve. It is designed to stimulate the natural neural activities of the vagus nerve to transmit signals up into the brain and down into the body.

Is VNS a suitable treatment option for my specific condition?

As we speak, the FDA has approved the sales and marketing of implantable vagus nerve stimulators to treat eligible patients with epilepsy or depression or who are recovering from a stroke and requiring rehabilitation therapy for the hand and arm.

Researchers worldwide are currently studying dozens of other conditions for the potential use of vagus nerve stimulators under FDA-sanctioned clinical trials, but these have not yet received approval for general clinical use in the United States. The list of these conditions is lengthy but includes rheumatoid arthritis, Crohn's disease, tinnitus, fibromyalgia, generalized anxiety disorder, irritable bowel syndrome, and others. You may be eligible to enlist as a volunteer in a clinical trial of vagus nerve stimulation for your specific condition, as Kelly Owens, Toney Kincaid, Pero Dragoje, and Nick Fournie did. You can find a complete list of these clinical trials on the clinicaltrials.gov website.

For a complete list of potential applications for vagus nerve stimulation and their trial and approval status, please see the appendix on page 263.

Is VNS for everyone?

While VNS holds promise for many conditions, it's not universally applicable. For some people, pre-existing conditions might influence the suitability of VNS therapy. Individuals with severe asthma or other breathing difficulties, like sleep apnea, may need to explore alternative treatment options due to potential worsening of these conditions with VNS. Certain heart problems, like abnormal rhythms or low heart rate, could also be negatively impacted by VNS, and a thorough evaluation by a cardiologist is required. VNS is also not recommended for individuals with only one vagus nerve (that is, one half of the paired structure, on only one side of the body).

How effective has VNS been in treating patients with a condition like mine?

The answer to this question depends on your specific clinical condition and the experiences of other patients who have been treated with vagus nerve stimulation for your health issue. In some conditions, there have been thousands of patients, but in others, the number of patients treated with VNS is less, making it harder to predict what VNS would do for you.

The largest cohort of treated patients is epilepsy, as hundreds of thousands of epilepsy patients have received vagus nerve stimulation over the past forty years. Approximately 50 percent or more of epilepsy patients have a reduced number of seizures, a significant clinical benefit, despite previously not deriving significant benefit from drugs. Fewer patients with depression have been treated with VNS, so the experience is less, but the clinical trial results that have been published indicate that about a third of patients experience improvement in depression, anxiety, and mood. For rheumatoid arthritis, there have only been several hundred patients treated worldwide with an implanted vagus nerve stimulator, but to date, the response rates for clinical improvement suggest a 50 percent or better positive outcome. Other conditions are being considered for vagus nerve stimulation—ask your physician for specifics regarding the statistics as they relate to your health issue.

But this science is still new enough that your physician may not know, which is why I wrote this book. I've tried to give you enough background so that as new information becomes available, you may be able to make better sense of what comes up on reliable medical websites in the future. And you may find the appendix to be helpful in pointing you to clinical trials for your specific clinical condition.

What is the procedure for implanting the VNS device?

The surgery can be performed under general anesthesia or local anesthesia, depending on your medical condition. Usually, these procedures do not require an overnight stay in the hospital and are performed in ambulatory surgery settings.

The early-generation devices require two incisions, one under the collarbone, where a pocket is formed under the skin to receive the pulse generator, a two-inch-diameter pacemaker-like device housing the battery and the computer. A second incision in the neck allows access to the vagus nerve, adjacent to your carotid artery. A wire lead is threaded between the two incisions, and one end of the wire is connected to the pulse generator. The other end is connected to the vagus nerve. As the years have passed and the technology has improved, vagus nerve stimulating devices continue to get smaller. Devices like the one I keep on my desk that's the size of a Tylenol capsule require only a single incision and can be implanted directly on the vagus nerve, eliminating the need for a separate pulse generator and for a second incision.

What about side effects?

With modern surgical techniques, the implantation of a VNS device is a relatively simple, largely uneventful procedure. Notably, less than 5 percent of patients experience postoperative side effects, including some that can occur with any surgery, such as wound infections or pain. Other infrequent complications include vocal cord weakness, neck pain, exertional shortness of breath (called *dyspnea*), headache, and others that you should review with your neurosurgeon. The removal or inactivation of the VNS device is exceedingly rare but may be required in the unlikely event of hardware malfunctions, infection, electrode fractures, or pacemaker failures. On the other hand, as many of the epilepsy patients discovered, side effects may also include improvements in your mood and outlook.

How and when does VNS therapy begin?

After you recover from surgery, usually one or two weeks later, your doctor will turn on the vagus nerve stimulation using a computer tablet or computerized magnetic wand specially designed to program the computer in your VNS device. On your first office visit, the device functions will be tested and therapy started, using settings that are specific to your condition. This is done by adjusting the frequency of pulses (usually 20–40 Hz), the duration of each pulse (usually 100–330 microseconds), and the current strength (usually 0.4–3.0 milliamperes). In the following days or weeks, depending on your specific diagnosis, you will return to your physician for assessments of your condition. At these visits, the current strength or other settings may be modified, a process that will be repeated as necessary after that.

How long does it typically take to see results from VNS therapy?

The clinical benefits of vagus nerve stimulation tend to increase with time. Some patients have significant improvements within the first few weeks, while many patients have noticed even better results beginning six months or more after the therapy begins.

Is VNS a permanent treatment, or can the device be removed in the future?

It is a permanent treatment, but the device can be turned off or removed if the clinical response is insufficient and if you wish to have it removed. Some devices may require a battery replacement because the battery life is on the order of ten to twelve years in the older devices. The newer-generation devices are unlikely to ever need a battery replacement since they are quite small and draw very little power, and because the battery technology has greatly improved.

Will I be able to feel the VNS device working, and what does it feel like?

That depends on the type of device that you have and the amount of electrical current necessary for your treatment. During VNS for epilepsy and depression, the devices typically deliver between 1 and 2.5 milliamperes of current, which can activate the nearby muscles in the neck and larynx. You might feel this as a vibration or tingling sensation. The early-generation devices are typically programmed to deliver current for 30 to 90 seconds at a time, followed by a 5-minute hiatus, around the clock. If you are speaking when the current comes on (during the 30 to 90-second period), it may cause your voice to vibrate or "change" a bit. But as research progresses in stimulating the vagus nerve to treat inflammation, we now find that the amount of current required to block cytokine release is much less. So, the risk of these muscle and voice side effects is limited to these few 5-minute or shorter periods in the course of a day. When these stimulations happen at night, some patients sleep right through them.

Can VNS therapy be used in conjunction with my current medications or other treatments?

Yes. For most conditions, the usual medications are continued for several days as the VNS device is turned on and your physician adjusts the strength of the electrical impulses. As your symptoms begin to improve, you may be able to reduce the dosages or quantities of your medications, but this should be done under the guidance of a physician because many medications require a gradual tapering before discontinuation.

How does VNS therapy impact daily activities or quality of life?

The majority of patients note that VNS has little or no effect on their daily activities. Certain limitations may apply, like avoiding high-intensity magnets. If you derive benefit for your condition from VNS therapy, then the major impact is improvement in your quality of life. Some patients who derive no significant clinical benefit for their condition nonetheless decide to retain the device because having a VNS leads them to have improvements in their mood and outlook.

What is the cost of VNS therapy, and is it covered by insurance?

The current cost of the VNS device itself can range from $30,000 to $50,000, and the cost of the surgery to implant the VNS device can range from $15,000 to $30,000. These numbers do vary according to geographic location, so the approximate total cost for the VNS device and surgery could be anywhere from $45,000 to $80,000. However, this is just a general estimate, and your costs will depend on your insurance and healthcare providers.

And of course, you should weigh these one-time costs against the alternatives.

Can VNS therapy be adjusted or turned off without surgery?

Yes. This is done by your physician using a specially designed wand or tablet.

How does VNS interact with specific brain regions related to different treatment targets (epilepsy, depression, etc.)?

Although there are known benefits of VNS in treating some conditions, we don't know the precise mechanisms that produce these benefits. Some proposed mechanisms for epilepsy include targeting the *seizure circuit* in structures like the brain's hippocampus, amygdala, and thalamus because stimulation of the vagus nerve sends signals up into these regions, modulating their activity and potentially preventing seizures from initiating or spreading. It is also possible that VNS influences the levels of neurotransmitters like GABA and glutamate, which play crucial roles in regulating brain excitability because increased GABA and decreased glutamate can help stabilize neuronal activity and reduce seizure susceptibility.

Some proposed mechanisms for depression include VNS stimulation influencing the brain's limbic structures, including the cingulate cortex and amygdala, potentially reducing their hyperactivity and alleviating depressive symptoms. VNS may also promote the production of brain-derived neurotrophic factor (BDNF), a protein crucial for neuronal growth and survival, and increased BDNF levels can enhance neuroplasticity and resilience in brain regions affected by depression, potentially contributing to mood improvement. Because VNS also blocks the production of inflammatory molecules known to cause depression in animals and humans, it is quite possible that the benefit observed using VNS to treat depression is attributable to VNS inhibiting the production of cytokines and other molecules underlying the onset of depression.

Can VNS help reduce chronic pain?

This is an area of active research because inflammation can enhance the sensitivity of neurons, called *nociceptors*, that transmit pain signals to the brain. Since VNS can block inflammation, it may be possible to use VNS to reduce pain. Furthermore, VNS can also modulate pain pathways in the brain stem and spinal cord.

Can VNS help with rheumatoid arthritis?

This is an area of active research, and early clinical trial results that have been published in the scientific and medical literature indicate that VNS can block inflammation and the major signs and symptoms of RA in some patients. As I write, a large multi-center U.S. randomized clinical trial is nearing completion under a "breakthrough" designation from the FDA. It is possible that by the time you read this, VNS will be a treatment option for some RA patients. These studies hold such significant promise for so many people that there are plans for additional studies of using VNS to reduce inflammation in other forms of arthritis and in other autoimmune and inflammatory conditions.

Can VNS help with inflammatory bowel disease?

This is another area of active research. Early clinical results in relatively small trials published in the scientific and medical literature indicate that VNS can block inflammation and the major signs and symptoms of IBD in some patients. Again, because these studies hold such significant promise for so many people, stay tuned for additional studies of using VNS to influence the gut-brain axis to reduce inflammation.

Could noninvasive methods, like auricular or transcutaneous approaches, or focused ultrasound, offer similar benefits to implants with fewer risks?

This is yet another area of active research. The challenge with these approaches is that, most commonly, the available devices deliver electrical current to the skin of the neck or ear, in the hopes of activating vagus nerve fibers. But there is no direct evidence that the electric pulses indeed pass through the skin and other layers of fat and muscle to reach the vagus nerve. It's possible that the electrical stimulation from these devices activates sensory neurons in the skin and other superficial tissues and muscles, whose signals are relayed to the brain, and that the brain may, in turn, relay outgoing signals through the vagus nerve. This is speculation, however, and has not been directly confirmed by experiment or clinical measurement of the vagus nerve. So, while these transcutaneous devices are somewhat simpler and certainly less expensive than surgically implanted devices, at this time the only way to know for sure that the vagus nerve is stimulated is to use an implanted device.

Focused ultrasound has been shown in clinical trials to directly stimulate the vagus nerve in the liver and spleen using noninvasive handheld probes, but unlike the auricular and transcutaneous electrodes and TENS units, these focused ultrasound devices are not yet available except as research prototypes in ongoing and planned additional clinical studies.

As understanding of the microbiome advances, can VNS be targeted to modulate the gut-brain axis for improved digestion and mental health?

Yes. This is an area of active research because of the relationship between the gut microbiome, appetite and feeding behavior, and metabolism. Ongoing research is exploring the relationship between signals derived from bacteria that are transmitted into the brain via the vagus nerve, and as knowledge of these signals improves, it may be possible to modify them with vagus nerve stimulating devices. Similarly, GLP-1 and other hormonelike molecules that control appetite and metabolism also activate signals in the vagus nerve. In the future, vagus nerve stimulating devices may offer an alternative approach to influence weight loss and regulate glucose and lipid levels in the bloodstream.

Could VNS be helpful in treating conditions like Alzheimer's or Parkinson's by supporting neural regeneration and synaptic function?

Yes. This is yet another area of active research and the subject of growing interest because of evidence that molecules produced in the gut can ascend in the vagus nerve to initiate the onset of

Alzheimer's-like, and Parkinson's-like, diseases in animal models. Furthermore, there is evidence in clinical trials and experimental models that inflammation can contribute to the progression of neurodegeneration in these conditions. So, researchers are studying whether it may be possible to inhibit the inflammatory responses in the brain, called *neuroinflammation*, using vagus nerve stimulation. The understanding of these mechanisms is in the very early days of scientific research.

Can VNS help me lose weight or regulate my blood sugar?

Early studies with mice of focused ultrasound directed at the vagus nerve in the liver suggest yes, though this technology is not yet FDA approved or available. VNS in the liver seems to have similar effects on appetite and metabolism as GLP-1 agonist drugs, such as Wegovy and Ozempic, that are FDA approved for treating diabetes and obesity.

What are these TENS units I've been seeing on social media, and how do they work?

TENS is short for *transcutaneous electrical nerve stimulation*. *Transcutaneous* means "across the skin," so broadly speaking, TENS units are electronic devices that, when strategically placed somewhere on your skin, are supposed to have some therapeutic effect by stimulating something under your skin. These devices are popular for what is commonly and even scientifically known as *transcutaneous auricular vagus nerve stimulation* (taVNS), when you place the device on your ear to stimulate—hopefully—the auricular branch of the vagus nerve. TENS units are relatively inexpensive and available online or over the counter without a prescription, but talk to your doctor before you use one.

TENS unit buyers should beware because the classification of a medical device is based on its intended use rather than its operational mechanism. This allows manufacturers to market devices as "noninvasive vagus nerve stimulators" through careful wording about the device's purpose without providing direct, or any, evidence proving that the stated mechanism is correct. There are many nerves in the ear and neck besides the vagus nerve branches, and there is no evidence that a TENS unit can selectively stimulate your vagus without also stimulating many other nerves in the neck and ear.

Biomedical engineers and neuroscientists uniformly agree that noninvasive skin electrode stimulation is not equivalent to applying an electrode onto the vagus nerve itself.

Implanting an electrode onto the vagus nerve or stimulating the nerve with focused ultrasound are the only ways we have today in humans to target specific mechanisms mediated by the vagus nerve. However, there is convincing, though incomplete and ongoing, clinical evidence that electrically stimulating the ear can inhibit inflammation and mediate other health benefits that may stem from increasing vagal tone.

Since VNS can help with depression, does it also have the potential to enhance cognitive function or mental resilience in healthy people?

This is another area of active research. Some evidence in people with conditions like epilepsy or depression has shown some cognitive improvements following VNS therapy, but these effects may be secondary to improvements in mood or seizure control. Studies in rodents have also shown that VNS can boost memory formation and consolidation, potentially through its influence on hippocampal activity. VNS has also been shown to enhance attentional focus in animal models, potentially by modulating activity in brain regions like the prefrontal cortex. But the proposed or experimental use of VNS as a specific method to modify cognition through these mechanisms is likely to face significant ethical and regulatory questions and issues.

Could VNS play a role in future brain-computer interfaces, directly linking our nervous system to technology?

Yes. VNS holds immense potential in the realm of brain-computer interfaces (BCIs), potentially serving as a crucial link between our nervous system and future technologies for many reasons. VNS can not only send signals to the brain but also receive information about its activity. This form of two-way communication could enable future BCIs to not only control devices but also gather real-time feedback from the brain, creating a more seamless and dynamic interaction.

From an ethical standpoint, how should we approach and regulate the potential for VNS to influence mood or behavior?

This question points to a complex ethical landscape. Here are some key considerations: Individuals have the right to make informed decisions about their own mental health and well-being, which includes the right to choose therapies like VNS, even if they hold the potential to influence mood or behavior. Certain individuals, like those with cognitive impairments or under undue pressure, may require additional safeguards to ensure informed consent and prevent manipulation. Clear guidelines prohibiting its use for nontherapeutic purposes are essential. Patients considering VNS therapy should be provided with clear and complete information about its potential effects on mood and behavior, both positive and negative. This includes potential side effects and long-term impacts.

VNS-based systems may collect and analyze brain activity and other physiological data. Robust data privacy measures and regulations are necessary to ensure its protection from unauthorized access, misuse, and discrimination. The algorithms used to interpret brain activity and modulate VNS stimulation should be transparent and subject to ethical review to prevent bias and unintended consequences. Establishing independent oversight bodies with expertise in neuroscience, ethics, and law is crucial to ensuring the responsible development and implementation of VNS technologies. As VNS technology evolves, ongoing ethical review processes are necessary to address new challenges and ensure its development upholds human rights and societal values. Understanding and addressing public concerns about the potential misuse of VNS technology, such as fears of creating a "mind control" society, is vital for building trust and responsible implementation.

How might VNS interact with artificial intelligence in future medical applications for personalized treatment protocols or real-time monitoring?

The potential for VNS to interact with AI in future medical applications holds promise for personalized treatment and real-time monitoring. AI analysis of individual brain data (through EEG or other means) may determine the optimal VNS stimulation parameters for a specific condition, mood state, and inflammation status. This could lead to tailored VNS protocols that are more effective and have fewer side effects. AI could analyze a patient's medical history, current symptoms, and VNS response to predict future episodes of depression, seizures, or other conditions. This information could then be used to adjust VNS settings proactively, potentially preventing or mitigating symptoms. Future closed-loop systems may enable AI to continuously monitor a patient's brain activity and to adjust VNS stimulation in real time based on what it detects in a dynamic and responsive treatment approach that constantly adapts to the patient's changing needs.

As technology advances, can we develop noninvasive VNS devices that are user-friendly and accessible for home use?

Yes. There are several research centers and companies working on developing focused ultrasound devices that are capable of directly stimulating the vagus nerve using ultrasound waves focused onto the vagus nerve in the liver and spleen. Clinical trials from the Feinstein Institutes and elsewhere indicate that focused ultrasound can inhibit the production of cytokines and other inflammatory molecules in humans and reduce glucose levels in patients with type 2 diabetes. An area of burgeoning research is based on combining AI-driven technology implanted in a hockey puck–sized device that can be worn on the skin to localize and stimulate the vagus nerve directly to treat chronic inflammatory conditions like rheumatoid arthritis and metabolic conditions like type 2 diabetes and obesity.

TaVNS researcher, taVNS user, and my friend Ulf Andersson, whom you met in chapter 8, jokes that you might consider wearing a helmet to the doctor's office if you plan to ask about vagus nerve stimulation. His point is that most doctors know little about the vagus nerve, and in his observation, some doctors "get grumpy" when they're confronted with their own lack of knowledge or with something they don't understand. Ulf told me about a former colleague of his who suffered from a severe autoimmune disease and came down with autoimmune hepatitis several years ago. While she waited for conventional treatment (high doses of corticosteroids) pending a biopsy that was hard to come by in the middle of the summer, she decided to try taVNS. She very quickly felt much better, and six weeks later, when she finally got the biopsy, her blood and tissue tests for liver inflammation were more or less normal.

Her doctor didn't believe her when she told him about her TENS device. He got mad and suggested she'd been "cheating" by treating herself with steroids.

I wrote this book for doctors as well as potential patients to offer a different way to have these potentially life-changing conversations. By admitting what we don't know yet, I aim to give credibility to what we do know already, share this knowledge with people who can use it, and encourage patients to advocate for themselves.

Coda:
The Clear and Present Future

Computer Chips, Not Medicines

The most reliable way to predict the future is to create it.
— DENNIS GABORN

What astonishes me . . . is the fact of finding myself here, and at this moment, deep in this life and not in any other. What stroke of chance has brought this about?
— SIMONE DE BEAUVOIR

We had an accident in the lab, and then, without any warning, I could see into the future," I said to Ralph Nappi, the vice-chair of the board of the Feinstein Institutes. We were lunching in style that day in 1998, seated in overstuffed chairs at the North Shore Steakhouse about a mile from my lab in Manhasset. I wasn't recounting a science fiction movie—like the one where a scientist gets teleported four centuries ahead by an accidental power surge into a time machine. I was updating Ralph about something that had just happened.

On a bar napkin, I sketched a stick figure with a brain inside a round head, vagus nerve trailing down into an egg-shaped body, where it connected to the outline of two lungs, a heart, and a spleen. "We put a molecule here, in the mouse's brain," I said, pointing. "We did this to stop inflammation in the mouse's brain. Then we accidentally administered an endotoxin here, in the abdomen." I gestured to the approximate middle of the oval. "We couldn't believe it. The experimental drug in the brain turned off inflammation in the body." I explained that since there wasn't enough of the molecule to get into the bloodstream, the brain and body had to be communicating in some other way. Looking up from my sketch, I saw that Ralph was smiling, so I continued.

"When we cut the vagus nerve, the anti-inflammatory effect vanished. *This* was the path of communication. Ralph, *we discovered that the vagus nerve slows down and inhibits inflammation.*"

Ralph nodded and acknowledged, "So that's why you started using electronic devices to stimulate the vagus nerve. It makes sense." Holding my gaze, he continued encouragingly, "Kevin, you need to patent that."

Convinced that this invention would create a new future for many people when computerized vagus nerve stimulators replace drugs, I did. I patented it.

Sometimes, futures are created out of decades of planning and work, like building a new island in the South China Sea, or sequencing the human genome. Other times, the creator stumbles into something. The history of science and medicine includes famous examples of both kinds of creations.

The future my colleagues and I are working on lies at the intersection of nerves and molecular medicine mechanisms and computer chips and electricity. These things converge—by chance and very much on purpose—to help people with therapies in the form of electrons instead of drugs, with fewer side effects and more efficacy. In this future we're creating, we find more ways to assist the body's own healing reflexes in maintaining homeostasis.

"The body" can sound clinical. What I mean is your body and my body and everybody. We're testing new theories of disease that not only build on the science that's come before but also see things differently. And we're turning these new ways of understanding illness into new solutions based in the vagus nerve. So, this clear and present future of bioelectronic medicine depends in no small part on the vagus nerve—yours and mine and everybody's.

Inventors predict the future by inventing it, and as I explained to Ralph even back in 1998, a new generation of vagus nerve stimulators has already been invented. I see a future when battery-powered vagus nerve stimulators, not drugs, treat inflammation in people, helping people whose drugs aren't currently helping, at all or enough. These same devices will be helpful for other good reasons, including providing an alternative to the high costs and terrible side effects of current drug therapies. In medicine and science, from the first patient to the hundreds of millions, innovation begins to reshape the world one person at a time, one clinical trial at a time, before changing the trajectory of human history to give us all hope that the end is in sight for a particular kind of suffering. Hope, as Kelly Owens dared to hope for herself and now wishes for everyone, "for health and a life free of pain." One life at a time.

A LIFE OF HEALTH

In the spring of 2017, Kelly Owens, whom you first met in chapter 1, was in Amsterdam with her husband, Sean. On their fifth wedding anniversary, they were sightseeing as much as their budget, and her painfully inflamed joints, would allow in between medical visits to prepare for surgery to implant a vagus nerve stimulator. Cobblestones were even harder on her body than the pavement she was used to in the United States, but she loved the look of those old streets, in a city where she'd never planned to be.

It had been about three years since Kelly first reached out to me and I'd suggested she keep an eye on clinical trials for Crohn's disease, apologizing that I couldn't do more at the time because that trial didn't exist yet. In the interim, Kelly's symptoms were so debilitating she had to leave her job as a teacher and give up some other dreams, like living in Hawaii.

She'd been in and out of hospitals for routine and acute treatments in the fifteen years since her diagnosis, including twelve straight days in early 2015 receiving fluids, nutrients, high doses of anti-inflammatory steroids, and morphine intravenously after the inflammation in her colon had gotten so bad that she was malnourished and emaciated. At the time, some well-meaning hospital staff complimented her on her weight loss. Kelly believed she was dying. She says that ever since inflammation had first announced itself when she was thirteen years old, her insides had felt like a loud, dysfunctional, door-slamming family (her metaphor). It was unpleasant, but at least they were alive and fighting. But eventually, her body seemed to be giving up.

"Every day, I woke up to my pain," Kelly recalls. "I didn't know what it was like to wake up without it. I opened my eyes and felt the heaviness of what it meant to live in my skin. My lower back throbbed, and my arms ached and tingled from my shoulder cuffs to the tips of each finger. Sometimes, I couldn't straighten my arms beyond a ninety-degree angle. Sometimes, my legs swelled up to the point of needing a pillow underneath my knees because I couldn't straighten them."

She was used to all that; she'd been living with it for fifteen years. Fifteen years of immobility from painful joints, often requiring a wheelchair or cane to ambulate, and depending on the day, her life limited even more by gastrointestinal distress. Fifteen years of prescription drug use, which failed to alleviate her suffering, and fifteen years experiencing their terrible side effects, all the while worrying about their black box warnings.

Kelly had tried every available biologic, DMARD, immunosuppressant, and (colon safe) anti-inflammatory medication, including prednisone, Remicade, Humira, Cimzia, Entyvio, Enbrel, methotrexate, Stelara, Simponi, Imuran, 6-MP, Flagyl, Pentasa, Entocort, Uceris, sulfasalazine, Asacol, Plaquenil, Celebrex, and Lialda. "None of them worked for too long, if at all," in her experience. She describes the effects of a few of them:

> *One of the worst injectables for me was Humira—I couldn't bear to give myself the injection. Sean had to do it because the medication burned so terribly upon injection that when I tried to do it myself, I couldn't leave the needle in for the full dose.*
>
> *Remicade made me sleep for days after the infusion.*
>
> *Injectable methotrexate left me in a chemo-brain-fog.*
>
> *Sulfasalazine left me nauseous and with a constant headache.*
>
> *Prednisone, oh, prednisone, a miracle drug as well as a haunted house of horrors. A drug that, when on super high doses, got rid*

of the inflammation but left me with the mania of a small army. The insomnia was torture, and when I would finally fall asleep, I'd find myself shooting straight up in bed only a few hours later, filled with dread and restless nerves that felt like ants were crawling in my veins.

And those were just the short-term side effects. Kelly says, "The world's worst balancing act is deciding what's worse: the disease symptoms or the possible long-term side effects, like lymphoma, rare cancers, osteoporosis, and more."

In Amsterdam, Kelly and Sean came across a small Banksy exhibition at the Moco Museum featuring his famous girl reaching for a heart-shaped red balloon, originally graffitied on London's Waterloo Bridge along with the words: THERE IS ALWAYS HOPE. With Kelly's implant surgery just weeks away, the artwork spoke to her.

Two days before the procedure, Sean asked Kelly what she planned to name her device. The pink cane her father had given her as a teenager was *Rosie*, after all, so it seemed fitting that her new bioelectronic *assistant* (as Sean then referred to it) get at least the same respect. She decided to call her vagus nerve stimulator *Murph* after a character in her favorite movie, Christopher Nolan's science fiction epic *Interstellar*, in which space travel is necessitated by an uninhabitable Earth. As Kelly saw it, Murph (played by Mackenzie Foy) is her father's (played by Matthew McConaughey) only hope.

Kelly's implant surgery was uncomplicated, and she left the hospital the same day, returning a few days later to have Murph turned on. She began to feel better almost immediately, and better and better over the following days, weeks, and months, during which she took careful notes. She was aware that if this thing worked, it would be "a big deal" not only for her but for everyone like her with an inflammatory disease who was also desperate for help. I am grateful to Kelly for allowing me to share these excerpts:

> *Day 1, July 6, 2017:*
> *I laid down on the table, and [the trial coordinator] waved a "wand" over Murph's battery to turn it on. The wand was attached to a tablet, and once she waved it over, she could control the settings of the device from the tablet and then wave it back over the device to "set" it. She turned up the frequency by increments of .25 milliamps until I could feel the stimulation—I finally felt it at 1.0, so we tried 1.25, and then 1.5, which felt like a little much, so we kept it at 1.25.*
>
> *That night, we made dinner and ate outside on the patio, enjoying the quietude of the gardens and the birds singing. When we got in bed at the end of the day, I curled into the covers comfortably and, after closing my eyes for a few seconds, realized that I hadn't needed any pain medication ever since stimulation was turned on twelve hours before.*
>
> *Day 3:*
> *Filled with emotion today, thinking about all of the people that have made it possible for us to be here. This place is beautiful.*
>
> *Day 3 of stimulation has my joints feeling the lightest they've ever felt.*
>
> *Day 6:*
> *As Sean and I were walking home from the train station, I felt an urge in my legs that I haven't felt in fifteen years—an urge to run. It felt like this long-lost freedom that I had once known that I forgot about, and I didn't know whether to laugh or cry or both.*

On day twelve since Murph was activated, Kelly was running late on the commute to her next clinical trial visit. Without thinking about it, she ran up a flight of stairs to catch her train. Two weeks later, she and Sean took a bus to Paris, where they spent the weekend walking, all day both days, she without slowing down or needing to be carried. When they flew back to New Jersey toward the end of the trial, a wheelchair was waiting as they got off the plane, confusing Kelly for a moment before she realized it was for someone else. She didn't need it.

Eight years later, Kelly has gone from someone who couldn't put on her own deodorant some days because her elbows were so swollen, or button her shirt, to someone who does fine woodworking, refinishing custom guitars. From someone who, after getting through a day of teaching (on days when her body would let her get up and go to work at all), had to lie on the couch with her legs propped up all evening to recover to someone who takes on too many projects and "doesn't know how to stop and rest." From someone who was dependent on drugs that, at best, helped suppress her inflammation a little bit to someone who doesn't need medications at all because her vagus nerve is being stimulated to do its job, putting the brakes on inflammation and keeping Kelly in a state of harmony and homeostasis.

For five years, Kelly worked for the Feinstein Institutes, sharing her story with the medical science community and fielding inquiries from people like herself who contacted us wanting to know about options. At conferences she attended with me, many of my physician and scientist colleagues were moved by Kelly's story. Others dismissed it as a placebo effect.

"That's a hell of a placebo," Kelly mused to me once after

someone had said this to her face. I told her that isn't how placebos work. The placebo effect is real, and people can become clinically better simply by receiving a sugar pill. But most placebo benefits do not last, typically fading away within three to four months. Kelly has been symptom- and treatment-free (besides her vagus nerve stimulating treatment) since 2017, far exceeding the usual duration of a placebo effect.

One CEO of a pharmaceutical company said to Kelly, in a tone that wasn't happy for her, "If your story is true, I'll be out of a job." To him I would say, My company's mission is to produce knowledge to cure disease. What's yours?

Kelly told me that many years before she got her vagus nerve stimulator, years before they'd heard of it, Sean once said to her, "Someday something is going to come along, and it's going to be like flipping a switch inside you to turn off what's wrong. But without extinguishing your light." Then he told her about all the electrical components and mechanisms of flipping a switch. Sean is a builder and contractor. And future-seer, apparently.

SEEING INTO THE FUTURE

To be clear, I am not a futurist or a soothsayer, but I am a neurosurgeon and scientist. There is a big difference between gazing into a crystal ball and counting the number of pond lilies to calculate the amount of time it will take for lily pads to cover the entire body of water. The former requires magic; the latter requires knowing the surface area of the pond, the pond area currently covered by lily pads, and the doubling time of the pond lilies. These things are all knowable. Armed with knowledge and data, it is then fairly simple to predict the future date at which the pond will be completely covered. Science produces the data that turns hope into inventions that create a new future.

Today I see data about a world where millions of people want better treatments for conditions caused by inflammation. They want treatments that do not have black box warnings, do not include the risk of bankruptcy as a side effect, and are effective at alleviating symptoms more than 40 percent of the time.

I see there is revolutionary new knowledge about the relationship between the vagus nerve, the brain, and the body's vital systems, including the immune system, which is unleashing powerful new insights that point the way to new therapies. The more we study these intersections and add to our knowledge about the operating mechanisms of the great nerve's protective and healing reflexes, the more we learn that using vagus nerve stimulation to treat inflammation and regulate homeostasis is not magic.

I see that we know how vagus nerve stimulation inhibits TNF and inflammation. We know how discrete neurons in the brain send their signals into the immune system of lab animals and humans. We know how these electrical signals are converted into chemical signals that change the behavior of white blood cells, converting them from angry inflammation producers to calm healing agents. And I see that new therapies based on this knowledge challenge the pharmaceutical industry status quo.

I see that forty years after my colleagues and I suggested using monoclonal anti-TNF antibodies, called *biologics*, to suppress inflammation in humans, we still do not understand how to prevent dangerous immunosuppression as a side effect. We do not know why anti-TNF confers significant benefit to only 40 percent of the patients. And we do not know why one biologic works in one patient while a similar biologic, which has the same chemical properties, fails to benefit the same patient. Someday, with more research, we may have answers to these and so many other questions about the current drug treatments for inflammation. But in the meantime, the knowledge base for using vagus nerve stimulation continues to grow. And after more than thirty years of clinical use, we know that, unlike the biologics, stimulating the vagus nerve does not cause potentially life-threatening immunosuppression.

I see that the clinical use of vagus nerve stimulators, which is based on the clinical experience of using cardiac pacemakers to treat millions of people for seventy years and counting, has, for the past three decades, been safely performed. Hundreds of thousands of patients with epilepsy, depression, and other conditions have received vagus nerve stimulator implants. This rich experience is now being applied to the development of vagus nerve stimulators to treat inflammation. Based on the number of emails I receive from patients seeking a vagus nerve stimulator, I believe there is an enormous pent-up demand for alternatives to the current standard biologic therapies.

I see and hear stories of desperation from patients who, like Kelly Owens, Toney Kincaid, Pero Dragoje, and Nick Fournie, need new treatments for their inflammatory diseases.

I see that vagus nerve stimulators have been implanted in patients with rheumatoid arthritis and inflammatory bowel disease, producing significant benefit in early clinical trials, and that larger clinical trials are presently underway and nearing completion. This research will hopefully pave the way for FDA approval of vagus nerve stimulation to treat inflammation and propel additional randomized clinical trials to assess whether these findings can be replicated in general populations.

I see patient activists like Kelly and Toney advocating for vagus nerve stimulation therapies as an effective and safer alternative to dangerous drugs. Beyond the conditions already studied (rheumatoid arthritis, inflammatory bowel disease, depression, epilepsy, and stroke rehabilitation), there's a pressing need to investigate vagus nerve stimulation in patients with other inflammatory conditions, including multiple sclerosis, long COVID, migraine, chronic pain, asthma, diabetes, obesity, hypertension, cardiovascular disease, POTS (postural orthostatic tachycardia syndrome), and cancer.

I see a significant increase in the number of inventors working on new, noninvasive technology that harnesses focused ultrasound to stimulate the vagus nerve. Such efforts are already underway, but they are in the early days. However, based on what my colleague Sangeeta and others have discovered, there is an accelerating pace of development focused on finding new ways to stimulate the vagus nerve to benefit future patients.

I see a growing number of people seeking to understand how to optimize their own vagus nerve activity, using do-it-yourself and at-home strategies. Some of these are time-honored ways to incorporate healthy practices, such as regular exercise and meditation, into everyday life. Some people combine these practices with wearables to track heart rate variability, and with over-the-counter TENS units to stimulate various nerves in their ear and neck in the hopes of boosting their vagal tone. Ongoing and planned clinical trials of these at-home modalities represent another major growth industry.

And I see, in my mind's eye, Grandpa Culotta, smiling down and urging us on, whispering, Let's do something about all this.

Acknowledgments

I wish to thank those who supported, enabled, and inspired the creation of this book. My wife, Tricia Tracey, whose boundless love, kind and generous spirit, and patient wisdom are a daily source of strength. Her expert eye and timely editing significantly improved the story and prose. My children and their spouses, Maureen, Patrick, Mary Bridget, Katherine, Jack, and Margaret, for tolerating my mental absences when my mind focused on this project, a frequent unfair distraction from their joyful, steadfast love and support. Stephanie Higgs, for her extraordinary writing, editing, and literary contributions that made this book possible. Her organizational brilliance, wordsmithing, recordkeeping, interviewing, smarts, and storytelling enlighten every page. My literary agent, Karen Murgolo, a first believer in this project, who read and edited countless proposals and drafts, then carried them forward to launch this book. Caroline Sutton, who led it through the gauntlet in the earliest stages of publication. My editor Lucia Watson, for sage guidance, advice, and the persistent focus on excellence now forever stitched into the book's spine. Richard Sloan, the early reader who, instead of spending his valuable time in his own laboratory studying the mind and the vagus nerve, donated it to this book, greatly enhancing it. And Bill Bruce, for the encouraging chats, over the fence *Tool Time* style, and for critically reading the early draft.

I am grateful to all my teachers and mentors, who taught me what they knew, showed me how to operate, and inspired me to discover and invent. Thanks to my most important teacher, my father, for choosing to face life with courage, wisdom, and a steadfast commitment to his family. My sincere indebtedness to T. Ross Kelly, Carl Franzblau, William McNary, G. Tom Shires, Stephen F. Lowry, Anthony Cerami, Murray Brennan, Russel Patterson, Carl Nathan, Hans Wigzell, and John Mountain. For collaborating, experimenting, and inventing with me, I thank my lab codirector Sangeeta Chavan and dearest friend, Ulf Andersson, along with Valentin Pavlov, Huan Yang, Eric Chang, Peder Olofsson, Jan Andersson, Mauricio Rosas-Ballina, Jared Huston, Haichao Wang, Luis Ulloa, Chris Czura, Richard Sloan, Ping Wang, Betty Diamond, Lionel Blanc, Peter Gregersen, Ben Lu, Yousef Al-Abed, Theo Zanos, Stavros Zanos, Timir Datta, Tomas Huerta, Timothy Billiar, John Eaton, Tak Mak, Yuman Fong, John Boockvar, Christine Metz, Barbara Sherry, Linc Moldawer, Marina Bianchi, Ona Bloom, Linda Watkins, Yoram Vodovotz, Jesse Roth, David Tuveson, Tony Zador, Mitchell Fink, Martine Rothblatt, and Michelle Iglesias.

I am grateful to Michael Dowling, for asking me to serve Northwell Health and the Feinstein Institutes, for encouraging me to write this book, and for giving me the opportunity to both lead the research programs and work with some of the most dedicated, compassionate, and caring healthcare professionals in the world. Special gratitude to Leonard and Susan Feinstein, whose

ACKNOWLEDGMENTS

vision, generosity, and belief in a nonprofit institute dedicated to producing knowledge to cure disease is the bedrock that this book is built upon. Thank you to the board members at Northwell Health and the Feinstein Institutes, who provided decades of support to the research that enabled the stories in the book to begin in the first place, long before others believed it could happen. I also thank my institutional benefactors, for supporting this work for decades, including the National Institute of General Medical Sciences of the National Institutes of Health, the Defense Advance Research Project Agency, the Gates Foundation, Northwell Health, SetPoint Medical, and United Therapeutics Inc. For countless hours of collaboration and friendship, and for cofounding SetPoint Medical, I thank my decades-long friend Shaw Warren. I am grateful to the Garvin family, who for generations have lovingly operated and maintained their unspoiled beachfront community, where I wrote much of this book.

Because this book, indeed all my efforts in science and medicine, stem from my fervent wish to alleviate suffering, I owe a debt of gratitude that can never be repaid to all the patients and physicians who took risks, personal and professional, to participate in early clinical trials, including Kelly Owens, Pero Dragoje, Toney Kincaid, Ante Bogut, Carlos Bravo-Iñiguez, Bill Bell, Ralph Zitnik, Richard Bucholz, Angela Crowley, David Chernoff, Paul-Peter Tak, Frieda Koopman, Sandra Milk, Jim Broderick, Mike Faltys, Ash Mehta, Allan Will, and hundreds of others. They took risk so future patients suffer less.

Appendix

A Bioelectronic Snapshot

The Current Status of Vagus Nerve Stimulation and Related Therapies for Various Conditions

As new evidence amasses, new applications for VNS and TENS line up for FDA approval, and research is ongoing. A search on March 23, 2024, for clinical trials using the term *vagus nerve stimulation* produced 449 studies. The following tables provide a snapshot of the clinical status as I write. You can search the website clinicaltrials.gov to see if there is an open clinical trial for your condition in real time.

Vagus Nerve Stimulation Using Implanted Devices

DISEASE OR CONDITION	INDICATIONS	FDA APPROVAL STATUS
Epilepsy	Used when seizures are not well controlled with medication	1997: FDA approved for adults and aged 12–17 2017: FDA approved for > 4 years old

Depression	Symptoms not controlled by multiple drugs or electroconvulsive therapy (ECT)	2005: FDA approved for adults
Rehabilitation after stroke	Combined with rehabilitation therapy to enhance function in hands and arms	2021: FDA approved for adults
Rheumatoid arthritis	Persistent symptoms not responding to multiple medications	FDA decision pending

FDA Approved for Transcutaneous Electrical Nerve Stimulation (TENS)

DISEASE OR CONDITION	DEVICE	FDA APPROVALS
Acute treatment of episodic cluster headache	gammaCore device applied to the neck	2017
Acute treatment of episodic migraine headache	gammaCore device applied to the neck	2018
Acute treatment of migraine headache	Nerivio device applied to upper arm	2019
Prevention or treatment of migraine headache	Cefaly device applied to forehead	2014

Acute treatment of migraine headache	Relivion device worn as a headband	2021
Opioid use disorder	NSS-2 Bridge device positioned behind the ear	2017
Opioid use disorder	Sparrow Ascent device applied to the ear	2023

Clinical Trials of TENS (abridged list)

CONDITION	STIMULATION SITE	CLINICALTRIALS.GOV
Epilepsy	Ear	NCT05031208
Tinnitus	Ear	NCT01176734
Mild cognitive impairment	Ear	NCT05514756
Type 2 diabetes mellitus	Ear	NCT02098447
Parkinson's disease	Ear	NCT04157621
Alzheimer's disease	Ear	NCT04908358
Fibromyalgia	Ear	NCT03180554
Major depressive disorder	Ear	NCT04467164

APPENDIX

Rheumatoid arthritis	Ear	NCT01569789
Hand osteoarthritis	Ear	NCT03919279
Irritable bowel syndrome	Ear	NCT05392439
Atrial fibrillation	Ear	NCT02548754
Acute stroke	Neck	NCT03733431
Post-traumatic stress	Neck	NCT05517304
Chronic pancreatitis	Neck	NCT03357029

Notes

Introduction: A New Frontier in Medicine

xi **the two hundred thousand or so:** You have two vagus nerves, one on each side of your neck, each comprised of approximately one hundred thousand nerve fibers. Henry Harland Hoffman and Harold Norman Schnitzlein, "The Numbers of Nerve Fibers in the Vagus Nerve of Man," *The Anatomical Record* 139, no. 3 (March 1961): 429–35, https://doi.org/10.1002/ar.1091390312.

Part One: Great Secrets

1 **illustration:** Image 444 from Andreae Vesalii Bruxellensis, *scholae medicorum Patauinae professoris, de humani corporis fabrica libri septem*, Jan Stephan van Calcar, illustrator (Basel, Switzerland: Joannes Oporinus, printer, 1543), https://www.loc.gov/resource/rbc0001.2023rosen0907/?sp=444&r=-0.769, -0.595,2.537,1.573,0.

1. How Electricity Could Replace Your Medications

3 **epigraph:** Charles Dickens, *Great Expectations* (Durham, NC: Duke Classics, 2012), chap. LIX, Kindle.
3 **The email caught my eye:** Kelly Owens, email to author, September 21, 2017.
3 **saw me on HuffPost Live:** "Dr. Kevin Tracey Explains Using Nerve Stimulation to Treat Arthritis," interview by Josh Zepps, May 30, 2014, HuffPost Live, video, 4:29, https://www.huffpost.com/entry/dr-kevin-tracey-explains-using-nerve-stimulation-to-treat-arthritis_n_5b4f9372e4b0cf38668f6ce0.
5 **a breakthrough designation:** "The Breakthrough Devices Program is a voluntary program for certain medical devices and device-led combination products that provide for more effective treatment or diagnosis of life-threatening or irreversibly debilitating diseases or conditions," from U.S.

Food & Drug Administration, "Breakthrough Devices Program: Guidance for Industry and Food and Drug Administration Staff," September 2023, https://www.fda.gov/regulatory-information/search-fda-guidance-documents/breakthrough-devices-program.

9   **Dr. William J. German:** William F. Collins and Lycurgus M. Davey, "William J. German, M.D., 1899–1981," *Journal of Neurosurgery* 54, no. 5 (May 1981): 571–72, https://doi.org/10.3171/jns.1981.54.5.0571.
10 **Survival is measured:** Soniya Mohammed, M. Dinesan, and T. Ajayakumar, "Survival and Quality of Life Analysis in Glioblastoma Multiforme with Adjuvant Chemoradiotherapy: A Retrospective Study," *Reports of Practical Oncology and Radiotherapy* 27, no. 6 (2022): 1026–36, https://doi.org/10.5603/RPOR.a2022.0113.
16 **When we published:** Kevin J. Tracey, "The Inflammatory Reflex," *Nature* 420, no. 6917 (December 2002): 853–59, https://doi.org/10.1038/nature01321; Lyudmila V. Borovikova et al., "Vagus Nerve Stimulation Attenuates the Systemic Inflammatory Response to Endotoxin," *Nature* 405, no. 6785 (May 25, 2000): 458–62, https://doi.org/10.1038/35013070.

2. The Great Nerve Reveals Itself

21 **epigraph:** Sir Charles Sherrington, *Man on His Nature* (Digital Library of India Item 2015.188837, 1940), 248.
22 **Vagal tone is the term:** Sylvain Laborde, Emma Mosley, and Julian F. Thayer, "Heart Rate Variability and Cardiac Vagal Tone in Psychophysiological Research—Recommendations for Experiment Planning, Data Analysis, and Data Reporting," *Frontiers in Psychology* 8 (February 19, 2017): 213, https://doi.org/10.3389/fpsyg.2017.00213; Fred Shaffer and Jay P. Ginsberg, "An Overview of Heart Rate Variability Metrics and Norms," *Frontiers in Public Health* 5 (September 28, 2017): 258, https://doi.org/10.3389/fpubh.2017.00258.
25 **his most famous experiment:** Charles G. Gross, "Galen and the Squealing Pig," *The Neuroscientist* 4, no. 3 (1998): 216–21, https://doi.org/10.1177/107385849800400317.
25 **Imagine the crowd:** Maud W. Gleason, "Shock and Awe: The Performance Dimension of Galen's Anatomy Demonstrations," Princeton/Stanford Working Papers in Classics Paper No. 010702 (January 2007), http://dx.doi.org/10.2139/ssrn.1427007.
26 **we had Aristotle's model:** Charles G. Gross, "Aristotle on the Brain," *The Neuroscientist* 1, no. 4 (July 1995): 245–50, https://doi.org/10.1177/107385849500100408.
30 **branches to the heart:** W. Bruce Fye, "Ernst, Wilhelm, and Eduard Weber," *Clinical Cardiology* 23, no. 9 (2000): 709–10, https://www.clinicalcardiology.org/briefs/200009briefs/cc23-709-profiles_wbr.html. Reporting on their experiments, the Weber brothers wrote, "If the medulla oblongata of a frog or the ends of the isolated vagus nerves are excited by the rotation of a fairly strong electro-magnetic machine, the heart suddenly stops beating, but at the end of excitation begins after a short interval to beat again: at

NOTES

first slowly and weakly, then gradually more strongly and more frequently until finally the original beat observed before excitation is restored."

32 **vagusstoff, which means:** Otto Loewi shared the 1936 Nobel Prize in Physiology or Medicine "for their discoveries relating to chemical transmission of nerve impulses." In his Nobel Lecture, he recounted his experiments using frogs to make the discovery that nerves convert electrical signals into chemical signals. "The Chemical Transmission of Nerve Action," Otto Loewi Nobel Lecture, The Nobel Prize, December 12, 1936, https://www.nobelprize.org/prizes/medicine/1936/loewi/lecture/.

33 **One study from Framingham:** William B. Kannel et al., "Heart Rate and Cardiovascular Mortality: The Framingham Study," *American Heart Journal* 113, no. 6 (June 1987): 1489–94, https://doi.org/10.1016/0002-8703(87)90666-1.

33 **larger study from France:** Athanase Benetos et al., "Influence of Heart Rate on Mortality in a French Population: Role of Age, Gender, and Blood Pressure," *Hypertension* 33, no. 1 (January 1999): 44–52, https://doi.org/10.1161/01.HYP.33.1.44.

36 **We sometimes gauge vagal tone:** Laborde, Mosley, and Thayer, "Heart Rate Variability," 213.

36 **One recent study that sounds:** Julia Shanks et al., "Cardiac Vagal Nerve Activity Increases During Exercise to Enhance Coronary Blood Flow," *Circulation Research* 133, no. 7 (August 29, 2023): 559–71, https://doi.org/10.1161/CIRCRESAHA.123.323017.

39 **One powerful tool:** Karl Deisseroth, "Optogenetics: 10 Years of Microbial Opsins in Neuroscience," *Nature Neuroscience* 18, no. 9 (2015): 1213–25, https://doi.org/10.1038/nn.4091.

39 **taking up these new tools:** Adam M. Kressel et al., "Identification of a Brainstem Locus That Inhibits Tumor Necrosis Factor." *Proceedings of the National Academy of Sciences* 117, no. 47 (November 9, 2020): 29803–10, https://doi.org/10.1073/pnas.2008213117.

39 **Researchers at Harvard:** Rui B. Chang et al., "Vagal Sensory Neuron Subtypes That Differentially Control Breathing," *Cell* 161, no. 3 (April 23, 2015): 622–33, https://doi.org/10.1016/j.cell.2015.03.022.

39 **four thousand to six thousand:** Nathalie Stakenborg et al., "Comparison Between the Cervical and Abdominal Vagus Nerves in Mice, Pigs, and Humans," *Neurogastroenterology & Motility* 32, no. 9 (May 31, 2020): e13889, https://doi.org/10.1111/nmo.13889.

41 **named then for the path:** Joseph Walsh, "Galen's Discovery and Promulgation of the Function of the Recurrent Laryngeal Nerve," *Annals of Medical History* 8, no. 2 (1926): 176–84, https://wellcomecollection.org/works/k88duuk8.

41 **how it was known:** Andrzej Żytkowski and Jerzy Walocha, "Anatomical Studies on Larynx and Voice Production in Historical Perspective," *Folia Medica Cracoviensia* (November 2020): 85-98, https://doi.org/10.24425/fmc.2020.135798; Charles Mayo Goss, "On Anatomy of Nerves by Galen of Pergamon," *American Journal of Anatomy* 118, no. 2 (March 1966): 327–35, https://doi.org/10.1002/aja.1001180202.

42 **nervus vagus in Latin:** Andrea Porzionato, Veronica Macchi, and Raffaele De

Caro, "The Role of Caspar Bartholin the Elder in the Evolution of the Terminology of the Cranial Nerves," *Annals of Anatomy—Anatomischer Anzeiger* 195, no. 1 (January 2013): 28–31, https://doi.org/10.1016/j.aanat.2012.04.007.

3. Your Body's Healing Reflexes

43 **epigraph:** Kelvin J. A. Davies, "Adaptive Homeostasis," *Molecular Aspects of Medicine* 49 (2016): 1–7, https://doi.org/10.1016/j.mam.2016.04.007. The full quote reads, "The constancy of the internal environment is the condition for free and independent life: the mechanism that makes it possible is that which assured the maintenance, within the internal environment, of all the conditions necessary for the life of the elements."

45 **"death note" in Janice's chart:** Because her death was from a home accident, state regulations required submission of specific forms and subsequent performance of an autopsy by the medical examiner.

48 **The technical term:** "Reciprocal inhibition" was one of many discoveries made by Sir Charles Sherrington, one of the two founders of neuroscience (the other is Ramón y Cajal). Sherrington won the Nobel Prize in Physiology or Medicine in 1932 for his life's work; https://www.nobelprize.org/prizes/medicine/1932/summary/.

50 **your coughing reflex:** This reflex was first illustrated by Ramón y Cajal, who received the Nobel Prize in Physiology or Medicine in 1906; https://www.nobelprize.org/prizes/medicine/1906/cajal/biographical/.

52 **Homeostasis emerges as something greater:** The notion of maintaining internal equilibrium was elucidated by French physiologist Claude Bernard in 1849. Bernard developed the concept of the internal environment, known as the *milieu intérieur*, and expressed the idea that the stability of this internal environment is crucial for enabling life to function freely and independently. This principle laid the foundation for the concept of *homeostasis*, a term introduced by Walter Cannon in his later works, e.g., Walter B. Cannon, *The Wisdom of the Body* (New York: W. W. Norton, 1932), 177–201 and W. B. Cannon, "Physiological Regulation of Normal States: Some Tentative Postulates Concerning Biological Homeostatics," in *A Charles Riches amis, ses collègues, ses élèves*, ed. A. Pettit (Paris: Les Éditions Médicales, 1926).

52 **heal:** *The Oxford Pocket Dictionary of Current English*, Encyclopedia.com, August 19, 2024, under "heal," https://www.encyclopedia.com/humanities/dictionaries-thesauruses-pictures-and-press-releases/heal-0.

52 **This short list:** Barry R. Komisaruk and Eleni Frangos, "Vagus Nerve Afferent Stimulation: Projection into the Brain, Reflexive Physiological, Perceptual, and Behavioral Responses, and Clinical Relevance," *Autonomic Neuroscience* 237 (January 2022): 102908, https://doi.org/10.1016/j.autneu.2021.102908.

53 **Their role is:** Cherly M. Heesch, "Reflexes That Control Cardiovascular Function," *Advances in Physiology Education* 277, no. 6 (December 1, 1999): S234–43, https://doi.org/10.1152/advances.1999.277.6.S234.

53 **Conversely, a decrease:** Sara L. Prescott and Stephen D. Liberles, "Internal

NOTES

Senses of the Vagus Nerve," *Neuron* 110, no. 4 (February 16, 2022): 579–99, https://doi.org/10.1016/j.neuron.2021.12.020.

53 **When you inhale deeply:** Prescott and Liberles, "Internal Senses."

54 **oxygenation of the blood:** B. H. Taha et al., "Respiratory Sinus Arrhythmia in Humans: An Obligatory Role for Vagal Feedback from the Lungs," *Journal of Applied Physiology* 78, no. 2 (February 1, 1995): 638–45, https://doi.org/10.1152/jappl.1995.78.2.638.

54 **signals that control various digestive:** Prescott and Liberles, "Internal Senses."

55 **Simultaneously, vagus and sympathetic:** Marie Aare Bentsen, Zaman Mirzadeh, and Michael W. Schwartz, "Revisiting How the Brain Senses Glucose—and Why," *Cell Metabolism* 29, no. 1 (January 8, 2019): 11–17, https://doi.org/10.1016/j.cmet.2018.11.001.

55 **The brain stem then coordinates:** W. Michael Panneton, "The Mammalian Diving Response: An Enigmatic Reflex to Preserve Life?," *Physiology* 28, no. 5 (September 1, 2013): 284–97, https://doi.org/10.1152/physiol.00020.2013.

56 **fundamentals of breathing:** Rui B. Chang et al., "Vagal Sensory Neuron Subtypes That Differentially Control Breathing," *Cell* 161, no. 3 (April 23, 2015): 622–33, https://doi.org/10.1016/j.cell.2015.03.022.

58 **"Inflammation and immunology must indeed":** Lewis Thomas, *The Lives of a Cell* (New York: Penguin, 1978), "Thoughts for a Countdown," Kindle.

59 **Eventually, my research would show:** Kevin J. Tracey et al., "Shock and Tissue Injury Induced by Recombinant Human Cachectin," *Science* 234, no. 4775 (October 24, 1986): 470–74, https://doi.org/10.1126/science.3764421.

60 ***tumor necrosis factor*, or simply TNF:** Kevin J. Tracey and Anthony Cerami, "Tumor Necrosis Factor, Other Cytokines and Disease," *Annual Review of Cell Biology* 9, no. 1 (November 1993): 317–43, https://doi.org/10.1146/annurev.cb.09.110193.001533 .

60 **Researchers were administering TNF:** E. A. Carswell et al., "An Endotoxin-Induced Serum Factor That Causes Necrosis of Tumors," *Proceedings of the National Academy of Sciences* 72, no. 9 (September 15, 1975): 3666–70, https://doi.org/10.1073/pnas.72.9.3666.

61 **When I saw the transformation:** Tracey et al., "Shock and Tissue."

62 **We crafted an antidote:** Kevin J. Tracey et al., "Anti-Cachectin/TNF Monoclonal Antibodies Prevent Septic Shock During Lethal Bacteraemia," *Nature* 330, no. 6149 (December 23, 1987): 662–64, https://doi.org/10.1038/330662a0. Monoclonal antibodies were then a seemingly miraculous new tool in modern science and medicine, first developed in 1975 by César Milstein and Georges Köhler using a breakthrough idea that garnered them the Nobel Prize. Monoclonal antibodies are produced by isolating and fusing a specific antibody-producing B-cell with a myeloma (cancer) cell, leading to an immortal cell line that produces large quantities of a single type of antibody. As will be explained below, monoclonal antibodies blocking cytokines, like anti-TNF, are today the basis for drugs to treat inflammation in millions of patients annually.

63 **Then we perfused:** Tracey et al., "Anti-Cachectin/TNF."

64 **In December of 1987:** Tracey et al., "Anti-Cachectin/TNF."
65 **named CNI-1493:** The name CNI-1493 was derived from Cytokine Networks Inc., the company that acquired the rights to do the clinical trials in the hopes of marketing the drug. Its number came from its place on a long list of molecules we were screening in the lab for maximal TNF-blocking activity. It was the 14th molecule in 1992 with positive activity, but we decided not to call it "1492," to avoid conflict with the year better known by Columbus "sailing the ocean blue," so instead we took poetic license and renamed it "1493."
65 **It proved to be exceptionally effective:** Stacey L. Oke and Kevin J. Tracey, "From CNI-1493 to the Immunological Homunculus: Physiology of the Inflammatory Reflex," *Journal of Leucocyte Biology* 83, no. 3 (March 2008): 512–17, https://doi.org/10.1189/jlb.0607363.
65 **It also worked in people:** Michael B. Atkins et al., "A Phase I Study of CNI-1493, an Inhibitor of Cytokine Release, in Combination with High-Dose Interleukin-2 in Patients with Renal Cancer and Melanoma," *Clinical Cancer Research* 7, no. 3 (March 2001): 486–92, PMID: 11297238.
66 **If the vagus nerve regulates:** Kevin J. Tracey, "The Inflammatory Reflex," *Nature* 420, no. 6917 (December 19, 2002): 853–59, https://doi.org/10.1038/nature01321.
67 **the results were in:** Lyudmila V. Borovikova et al., "Vagus Nerve Stimulation Attenuates the Systemic Inflammatory Response to Endotoxin." *Nature* 405, no. 6785 (May 25, 2000): 458–62, https://doi.org/10.1038/35013070.
68 *Immunologists are often asked*: R. A. Good, "Foreword: Interactions of the Body's Major Networks," in *Psychoneuroimmunology*, ed. Robert Ader (Orlando: Academic Press, 1981), xvii–xix.
69 **an entirely new field of medicine:** Kevin J. Tracey, "Shock Medicine," *Scientific American* 312, no. 3 (February 2015): 28–35, https://www.researchgate.net/publication/276586171_Shock_Medicine.

Part Two: Great Interventions

71 **art:** Santiago Ramón y Cajal, "Scheme to Explain the Mechanism of Vomiting and Coughing" (Madrid: CSIC [Spanish National Research Council], 1904), https://www.csic.es/en/csic.

4. The Path to Stimulation and Early Experiments with Epilepsy

73 **epigraph:** Mary Wollstonecraft Shelley, *Frankenstein; Or, The Modern Prometheus* (Durham, NC: Duke Classics, 2012).
74 **neurons propagate distinct spikes:** A. L. Hodgkin and A. F. Huxley, "A Quantitative Description of Membrane Current and its Application to Conduction and Excitation in Nerve," *The Journal of Physiology* 117, no. 4 (August 28, 1952): 500–44, https://doi.org/10.1113/jphysiol.1952.sp004764; E. R Kandel, J. H Schwartz, and T. M. Jessell, *Principles of Neural Science*, 4th ed. (New York: McGraw-Hill, 2000).

NOTES

75 **We do this with an electrical:** Pegah Afra et al., "Evolution of the Vagus Nerve Stimulation (VNS) Therapy System Technology for Drug-Resistant Epilepsy," *Frontiers in Medical Technology* 3 (August 25, 2021): 696543, https://doi.org/10.3389/fmedt.2021.696543.

75 **tempo of a vagus nerve:** Afra et al., "Evolution of the Vagus Nerve Stimulation (VNS)."

77 **You might be surprised:** Dènahin Hinnoutondji Toffa et al., "Learnings from 30 Years of Reported Efficacy and Safety of Vagus Nerve Stimulation (VNS) for Epilepsy Treatment: A Critical Review," *Seizure* 83 (October 10, 2020): 104–23, https://doi.org/10.1016/j.seizure.2020.09.027.

77 **In a late-1800s innovation:** Douglas J. Lanska, "J. L. Corning and Vagal Nerve Stimulation for Seizures in the 1880s," *Neurology* 58, no. 3 (February 12, 2002): 452–59, https://doi.org/10.1212/WNL.58.3.452.

78 **"The spindles disappeared":** A. Zanchetti, S. C. Wang, and G. Moruzzi, "The Effect of Vagal Afferent Stimulation on the EEG Pattern of the Cat," *Electroencephalography and Clinical Neurophysiology* 4, no. 3 (August 1952): 357–61, https://doi.org/10.1016/0013-4694(52)90064-3.

79 **Thanks to Zoll:** Paul M. Zoll, "Resuscitation of the Heart in Ventricular Standstill by External Electric Stimulation," *New England Journal of Medicine* 247, no. 20 (November 13, 1952): 768–71, https://doi.org/10.1056/NEJM195211132472005.

80 **Then came the transistor:** Developed at Bell Labs in Murray Hill, NJ, by John Bardeen, Walter Brattain, and William Shockley, the transistor revolutionized electronics and reverberated across other fields, including neuroscience, medical devices, and vagus nerve stimulators.

81 **During a three-hour power failure:** Oscar Aquilina, "A Brief History of Cardiac Pacing," *Images in Paediatric Cardiology* 8, no. 2 (April–June 2006): 17–81, PMCID: PMC3232561.

81 **The colleague, Earl Bakken:** Aquilina, "A Brief History."

81 **an article in *Popular Electronics*:** In the April 1956 issue.

81 **In his autobiography:** Aquilina, "A Brief History"; Earl E. Bakken, *One Man's Full Life* (Minneapolis: Medtronic Inc., 1999), 50.

81 **a surgeon in Stockholm:** Lawrence K. Altman, "Arne H. W. Larsson, 86; Had First Internal Pacemaker," *The New York Times*, January 18, 2002, https://www.nytimes.com/2002/01/18/world/arne-h-w-larsson-86-had-first-internal-pacemaker.html. Åke Senning, a surgeon leading the Department of Thoracic Surgery at Karolinska Hospital, performed the implant surgery.

82 **millions of such devices:** "Medtronic Reports Full Year and Fourth Quarter Fiscal 2023 Financial Results; Announces Dividend Increase," Medtronic, May 25, 2023, https://news.medtronic.com/2023-05-25-Medtronic-reports-full-year-and-fourth-quarter-fiscal-2023-financial-results-announces-dividend-increase.

82 **neuroscientist Jacob Zabara:** Barnaby J. Feder, "Battle Lines in Treating Depression," *The New York Times*, September 10, 2006, https://www.nytimes.com/2006/09/10/business/yourmoney/10cyber.html.

82 **He was aware of Zanchetti's:** Jacob Zabara, neurocybernetic prosthesis,

U.S. Patent 4,702,254, filed December 30, 1985, and issued October 27, 1987, https://patents.justia.com/patent/4702254.

82 **He teamed up:** Jacob Zabara cofounded Cyberonics with Reese S. Terry Jr.

83 **Toney Kincaid became the first:** Andrew Fuller, "An Electronic Victory Against Epilepsy," *Invention & Technology*, Summer 2000, https://www.inventionandtech.com/content/electronic-victory-against-epilepsy-1.

83 **In his pivotal work:** Hippocrates, "On the Sacred Disease," trans. Francis Adams, 400 BCE, https://classics.mit.edu/Hippocrates/sacred.html.

83 **We now understand that epilepsy's:** Orrin Devinsky et al., "Epilepsy," *Nature Reviews Disease Primers* 3, no. 18024 (May 3, 2018): 1–24, https://doi.org/10.1038/nrdp.2018.24.

85 **two thousand years after Hippocrates:** Christian M. Kaculini, Amelia J. Tate-Looney, and Ali Seifi, "The History of Epilepsy: From Ancient Mystery to Modern Misconception," *Cureus* 13, no. 3 (March 17, 2021), https://doi.org/10.7759/cureus.13953.

85 **characteristic electrical discharges:** Devinsky et al., "Epilepsy."

85 **With over twenty available ASMs:** Devinsky et al., "Epilepsy."

87 **care of his neurologist:** Dr. J. Kiffin Penry

87 **his first seizure-free day:** Antoinette Kerr, "Epilepsy Awareness Month: Locals like Kincaid Fight for Awareness and Legislation," *Davidson Local*, November 13 (year not noted), https://www.davidsonlocal.com/news/epilepsy-awareness-month-toney-kincaid-fights-for-awareness.

87 **Their mission is:** Antoinette G. Kerr, "Epilepsy Awareness Advocates and Others Raise Awareness About Disabilities," *Davidson Local*, December 4, 2023, https://epilepsyassocaitionofnc.org/f/toney-kincaid-and-members-of-the-epilepsy-association-of-north-ca.

87 **In the initial study:** J. Kiffin Penry and J. Christine Dean, "Prevention of Intractable Partial Seizures by Intermittent Vagal Stimulation in Humans: Preliminary Results," *Epilepsia* 31 (June 1990): S40–43, https://doi.org/10.1111/j.1528-1157.1990.tb05848.x.

88 **One patient experienced:** Penry and Dean, "Prevention of Intractable Partial Seizures."

88 **Building on the pilot studies:** E. Ben-Menachem et al., "Vagus Nerve Stimulation for Treatment of Partial Seizures: 1. A Controlled Study of Effect on Seizures," *Epilepsia* 35, no. 3 (May–June 1994): 616–26, https://doi.org/10.1111/j.1528-1157.1994.tb02482.x.

89 **Side effects are rare:** Toffa et al., "Learnings from 30 Years."

90 **Years of research and clinical:** Kristl Vonck et al., "The Mechanism of Action of Vagus Nerve Stimulation for Refractory Epilepsy: The Current Status," *Journal of Clinical Neurophysiology* 18, no. 5 (September 2001): 394–401, https://doi.org/10.1097/00004691-200109000-00002.

90 **it suppresses inflammation:** Yue Wang et al., "Vagus Nerve Stimulation in Brain Diseases: Therapeutic Applications and Biological Mechanisms," *Neuroscience & Biobehavioral Reviews* 127 (August 2021): 37–53, https://doi.org/10.1016/j.neubiorev.2021.04.018.

90 **VNS therapy may sway:** Francesco Marrosu et al., "Correlation Between

GABA$_A$ Receptor Density and Vagus Nerve Stimulation in Individuals with Drug-Resistant Partial Epilepsy," *Epilepsy Research* 55, no. 1–2 (June–July 2003): 59–70, https://doi.org/10.1016/S0920-1211(03)00107-4.

90 **triggers the release of endogenous:** Yi-Ting Fang et al., "Neuroimmunomodulation of Vagus Nerve Stimulation and the Therapeutic Implications," *Frontiers in Aging Neuroscience* 15 (July 5, 2023), https://doi.org/10.3389/fnagi.2023.1173987; Monja P. Neuser et al., "Vagus Nerve Stimulation Boosts the Drive to Work for Rewards," *Nature Communications* 11, no. 1 (July 16, 2020): 3555, https://doi.org/10.1038/s41467-020-17344-9; Eugenijus Kaniusas et al., "Current Directions in the Auricular Vagus Nerve Stimulation I–A Physiological Perspective," *Frontiers in Neuroscience* 13 (2019): 464811, https://doi.org/10.3389/fnins.2019.00854.

90 **Still another theory:** Kaniusas et al., "Current Directions"; Yue Wang et al., "Vagus Nerve Stimulation in Brain Diseases: Therapeutic Applications and Biological Mechanisms," *Neuroscience & Biobehavioral Reviews* 127 (August 2021): 37–53, https://doi.org/10.1016/j.neubiorev.2021.04.018.

91 **may foster neural plasticity:** Wang et al., "Vagus Nerve Stimulation."

5. Rebalancing Inflammation

93 **epigraph:** Lewis Thomas, *Late Night Thoughts on Listening to Mahler's Ninth Symphony* (New York: Penguin Books, 1995), 44.

93 **Although their death certificates:** Roma Pahwa, Amandeep Goyal, and Ishwarlal Jialal, *Chronic Inflammation*, updated August 7, 2023 (Treasure Island, FL: StatPearls Publishing, January 2024), https://www.ncbi.nlm.nih.gov/books/NBK493173/; "The Top 10 Causes of Death," Newsroom, World Health Organization, December 9, 2020, https://www.who.int/news-room/fact-sheets/detail/the-top-10-causes-of-death.

95 **The FDA approval process:** "The Drug Development Process," U.S. Food & Drug Administration, content current as of January 4, 2018, https://www.fda.gov/patients/learn-about-drug-and-device-approvals/drug-development-process.

96 **lead clinical targets:** We chose RA and IBD primarily because they are commonly managed using monoclonal, anti-TNF antibodies.

96 **my publication in *Nature*:** Kevin J. Tracey et al., "Anti-cachectin/TNF Monoclonal Antibodies Prevent Septic Shock During Lethal Bacteraemia," *Nature* 330, no. 6149 (December 23, 1987): 662–64, https://doi.org/10.1038/330662a0.

96 **Biological anti-inflammatory drugs:** Yasmine El Abd et al., "Mini-Review: The Market Growth of Diagnostic and Therapeutic Monoclonal Antibodies–SARS CoV-2 as an Example," *Human Antibodies* 30, no. 1 (February 18, 2022): 15–24, https://doi.org/10.3233/HAB-211513.

97 **We began calling this intersection:** Kevin J. Tracey, "Shock Medicine," *Scientific American* 312, no. 3 (February 2015): 28–35, https://www.researchgate.net/publication/276586171_Shock_Medicine.

98 **"Joseph Meister lived":** This anecdote about Pasteur's supposed epitaph

request seems to be popular in faith-based literature. I have not found a proper historical source.

98 **Jonas Salk invented:** David M. Oshinsky, *Polio: An American Story* (Oxford: Oxford University Press, 2005); "History of the Polio Vaccine," Newsroom, World Health Organization, accessed March 22, 2024, https://www.who.int/news-room/spotlight/history-of-vaccination/history-of-polio-vaccination#:~:text=Salk%20tested%20his%20experimental%20killed,licensed%20on%20the%20same%20day.

99 **Named from the Greek words:** Josef S. Smolen, Daniel Aletaha, and Iain B. McInnes, "Rheumatoid Arthritis," *The Lancet* 388, no. 10055 (October 22, 2016): 2023–38, https://doi.org/10.1016/S0140-6736(16)30173-8.

99 **Rheumatoid arthritis affects:** Marita Cross et al., "The Global Burden of Rheumatoid Arthritis: Estimates from the Global Burden of Disease 2010 Study," *Annals of the Rheumatic Diseases* 73, no. 7 (July 2014): 1316–22, https://doi.org/10.1136/annrheumdis-2013-204627.

99 **Medications and pills:** Smolen, Aletaha, and McInnes, "Rheumatoid Arthritis."

100 **Another class of anti-inflammatory drugs:** Yoshiya Tanaka, "Recent Progress in Treatments of Rheumatoid Arthritis: An Overview of Developments in Biologics and Small Molecules, and Remaining Unmet Needs," *Rheumatology* 60, Issue Supplement 6 (November 2021): vii2–20, https://doi.org/10.1093/rheumatology/keab609.

100 **They also have high costs:** Tanaka, "Recent Progress."

102 **Richard brought an:** Richard Bucholz built in his basement the first prototype of a navigational system that subsequently became the StealthStation. This device is now marketed by Medtronic and has become the standard of care for essentially all cranial surgery, and increasingly for spinal and ENT surgery.

102 **At SetPoint, Yaakov had discovered:** Through experiments involving sophisticated nerve-stimulating and recording electrodes in mice and rats, Yaakov Levine demonstrated that a mere few hundred microamps of current applied to the vagus nerve effectively curbed the production of inflammatory cytokines, tenfold less than the amount of current used in the typical settings to treat epilepsy. Yaakov's finding suggested that the risk of side effects should be significantly lessened because the low amount of current is sufficient to activate only a fraction of the one hundred thousand individual nerve fibers in one side of the vagus nerve. Crucially, Yaakov discovered that the fraction that is activated includes the fibers responsible for inhibiting inflammation.

102 **he modified the factory settings:** The chief engineer at SetPoint Medical, Mike Faltys, a brilliant inventor, biomedical engineer, and kind friend, reprogrammed the device for rheumatoid arthritis based on Yaakov's discovery.

106 **In addition to the first patient:** Frieda A. Koopman et al., "Vagus Nerve Stimulation Inhibits Cytokine Production and Attenuates Disease Severity in Rheumatoid Arthritis," *Proceedings of the National Academy of Sciences* 113, no. 29 (July 5, 2016): 8284–89, https://doi.org/10.1073/pnas.1605635113.

Ante Bogut recalls his hospital had two of the seventeen patients in the RA trial, Pero Dragoje and a woman who also improved but whose baseline was low. Her improvement wasn't as dramatic as Pero's since her symptoms weren't as severe to begin with. Sadly, she died of COVID-19 in 2020.

107 **Even more important for:** To assess the severity of RA clinically, rheumatologists use a scoring system called the *DAS28-CRP* score, a standard numerical measure of the disease activity as a composite of counting the number of swollen painful joints and testing the blood for inflammation by measuring the levels of C-reactive protein, or just CRP. Vagus nerve stimulation caused a significant improvement in DAS28-CRP values from their baseline levels. Together with the TNF measurements, these results indicated that vagus nerve stimulation has a significant effect on controlling TNF production and the severity of rheumatoid arthritis, which is comparable, if not superior, to the results obtained using biologics that require weekly or monthly injections.

107 **Five years later, in:** Frieda A. Koopman et al., "Vagus Nerve Stimulation"; The last author of the article, the senior position in a peer-reviewed publication like this, was Paul Peter Tak, a physician-scientist in rheumatology from the Academic Medical Center in Amsterdam. The first author was Frieda Koopman, his former student. In 2009, Paul Peter Tak and his colleagues at the University of Amsterdam had made an interesting observation that inflammatory cells obtained from the joints of patients with rheumatoid arthritis express the receptor, or docking site on cells, for acetylcholine. They observed that the subtype of this receptor, called *alpha7 nicotinic acetylcholine receptor*, was the same receptor subtype that my lab had described in a 2003 publication in *Nature* after we discovered it is the key receptor that turns off the production of cytokines. Like a lock and key, putting the acetylcholine key into an alpha7 lock is a crucial anti-inflammatory mechanism that blocks inflammation. Together, these intriguing associations meant that vagus nerve stimulation could activate acetylcholine, which in turn switches off the inflammatory cells in the joints to stop the pain and swelling in rheumatoid arthritis patients. Tak's expertise was invaluable in shaping the trial's design and methodology.

107 **Called the *RESET-RA study*:** "Vagus Nerve Stimulation for Moderate to Severe Rheumatoid Arthritis (RESET-RA)," National Library of Medicine, last updated February 28, 2024, https://clinicaltrials.gov/study/NCT04539964.

108 **Then, by the time:** The press release went out July 10, 2024, e.g., https://www.businesswire.com/news/home/20240710182145/en/SetPoint-Medical-Announces-Positive-Topline-Results-from-Landmark-RESET-RA-Study-Evaluating-Neuroimmune-Modulation-for-the-Treatment-of-Rheumatoid-Arthritis.

108 **74 percent of rheumatoid arthritis:** "CreakyJoints Study Finds 74 Percent of Rheumatoid Arthritis Patients Dissatisfied with Treatment," *Businesswire*, September 10, 2019, https://www.businesswire.com/news/home/201909 10005998/en/CreakyJoints-Study-Finds-74-Percent-of-Rheumatoid-Arthritis-Patients-Dissatisfied-with-Treatment.

109 **Named for Dr. Burrill Crohn:** Burrill B. Crohn, Leon Ginzburg, and Gordon D. Oppenheimer, "Regional Ileitis: A Pathologic and Clinical Entity," *Journal of the American Medical Association* 99, no. 16 (October 15, 1932): 1323–29.

110 **Since Crohn's first description:** Daniel C. Baumgart and William J. Sandborn, "Crohn's disease," *The Lancet* 380, no. 9853 (November 3, 2012): 1590–1605, https://www.thelancet.com/article/S0140-6736(12)60026-9/fulltext.

111 **This clinical study, which:** Geert D'Haens et al., "Neuroimmune Modulation Through Vagus Nerve Stimulation Reduces Inflammatory Activity in Crohn's Disease Patients: A Prospective Open-Label Study," *Journal of Crohn's and Colitis* 17, no. 12 (December 2023): 1897–1909, https://doi.org/10.1093/ecco-jcc/jjad151.

112 **But after decades:** Dènahin Hinnoutondji Toffa et al., "Learnings from 30 Years of Reported Efficacy and Safety of Vagus Nerve Stimulation (VNS) for Epilepsy Treatment: A Critical Review," *Seizure* 83 (October 10, 2020): 104–23, https://doi.org/10.1016/j.seizure.2020.09.027.

113 **anti-TNF and other biologic agents:** Alvaro San-Juan-Rodriguez et al., "Trends in List Prices, Net Prices, and Discounts of Self-Administered Injectable Tumor Necrosis Factor Inhibitors," *Journal of Managed Care & Specialty Pharmacy* 27, no. 1 (December 30, 2021): 112–17, https://doi.org/10.18553/jmcp.2021.27.1.112.

113 **I keep a vagus nerve:** Michael Behar, "Can the Nervous System Be Hacked?" *The New York Times Magazine*, May 23, 2014, https://www.nytimes.com/2014/05/25/magazine/can-the-nervous-system-be-hacked.html.

6. Beyond Medication: A Healing Reflex for Depression

115 **epigraph:** Kay Redfield Jamison, *An Unquiet Mind: A Memoir of Moods and Madness* (New York: Knopf, 1995), 174.

115 **"A lot of hatred":** Kathleen Berger, producer, "Depressed Patients See Quality of Life Improve with Implanted Nerve Stimulation Device," HEC Science & Technology, YouTube video, 7:07, August 5, 2019, https://youtu.be/7SZ2obz8cUg?si=8Rx1KFtuK5Z-12Zg.

116 **It is becoming more so:** "Depressive Disorder (Depression)," World Health Organization, March 31, 2023, https://www.who.int/news-room/fact-sheets/detail/depression.

116 **As one of nearly:** Charles R. Conway et al., "Chronic Vagus Nerve Stimulation Significantly Improves Quality of Life in Treatment-Resistant Major Depression," *The Journal of Clinical Psychiatry* 79, no. 5 (August 21, 2018): 22269, https://doi.org/10.4088/JCP.18m12178.

117 **a pacemaker for the brain:** Berger, "Depressed Patients."

117 **The participants in this study:** Conway et al., "Chronic Vagus Nerve Stimulation."

117 **It found that when they:** Jim Dryden, "Depressed Patients See Quality of Life Improve with Nerve Stimulation," WashU Medicine, August 21, 2018,

https://medicine.wustl.edu/news/depressed-patients-see-quality-of-life-improve-with-nerve-stimulation-therapy/.

119 **it improves their mood:** Dènahin Hinnoutondji Toffa et al., "Learnings from 30 Years of Reported Efficacy and Safety of Vagus Nerve Stimulation (VNS) for Epilepsy Treatment: A Critical Review," *Seizure* 83 (October 10, 2020): 104–23, https://doi.org/10.1016/j.seizure.2020.09.027.

120 **began to use language:** Steven D. Hollon et al., "Evolutionary Theory and the Treatment of Depression: It Is All About the Squids and the Sea Bass," *Behaviour Research and Therapy* 143 (August 2021): 103849, https://doi.org/10.1016/j.brat.2021.103849.

121 **the vagus nerve is crucial:** Barry R. Komisaruk and Eleni Frangos, "Vagus Nerve Afferent Stimulation: Projection into the Brain, Reflexive Physiological, Perceptual, and Behavioral Responses, and Clinical Relevance," *Autonomic Neuroscience* 237 (January 2022): 102908, https://doi.org/10.1016/j.autneu.2021.102908.

122 **HRV tends to be:** Andrew H. Kemp et al., "Impact of Depression and Antidepressant Treatment on Heart Rate Variability: A Review and Meta-Analysis," *Biological Psychiatry* 67, no. 11 (June 1, 2010): 1067–74, https://doi.org/10.1016/j.biopsych.2009.12.012; Dominique L. Musselman, Dwight L. Evans, and Charles B. Nemeroff, "The Relationship of Depression to Cardiovascular Disease: Epidemiology, Biology, and Treatment," *Archives of General Psychiatry* 55, no. 7 (July 1998): 580–92, https://www.academia.edu/99347913/The_Relationship_of_Depression_to_Cardiovascular_Disease.

123 **The catecholamines released:** Michael A. Flierl et al., "Upregulation of Phagocyte-Derived Catecholamines Augments the Acute Inflammatory Response," *PLOS One* 4, no. 2 (February 12, 2009): e4414, https://doi.org/10.1371/journal.pone.0004414.

123 **This state, called *trained immunity*:** Charlotte D.C.C. van der Heijden et al. "Catecholamines Induce Trained Immunity in Monocytes in Vitro and in Vivo," *Circulation Research* 127, no. 2 (April 2020): 269–83, https://doi.org/10.1161/CIRCRESAHA.119.315800.

123 **Controlled experiments with human:** Julia A. Penatzer et al., "Salivary Cytokines as a Biomarker of Social Stress in a Mock Rescue Mission," *Brain, Behavior, & Immunity-Health* 4 (April 2020): 100068, https://doi.org/10.1016/j.bbih.2020.100068; Yvette Z. Szabo, Danica C. Slavish, and Jennifer E. Graham-Engeland, "The Effect of Acute Stress on Salivary Markers of Inflammation: A Systematic Review and Meta-Analysis," *Brain, Behavior, and Immunity* 88 (August 2020): 887–900, https://doi.org/10.1016/j.bbi.2020.04.078.

123 **Since psychological stressors:** Anna L. Marsland et al., "The Effects of Acute Psychological Stress on Circulating and Stimulated Inflammatory Markers: A Systematic Review and Meta-Analysis," *Brain, Behavior, and Immunity* 64 (2017): 208–19.

124 **Although many studies support:** Emanuele F. Osimo et al., "Inflammatory Markers in Depression: A Meta-Analysis of Mean Differences and Variability

in 5,166 Patients and 5,083 Controls," *Brain, Behavior, and Immunity* 87 (July 2020): 901–9, https://doi.org/10.1016/j.bbi.2020.02.010.

125 **called *sickness syndrome*:** Steven F. Maier and Linda R. Watkins, "Immune-to-Central Nervous System Communication and Its Role in Modulating Pain and Cognition: Implications for Cancer and Cancer Treatment," *Brain, Behavior, and Immunity* 17, no. 1 (February 15, 2003): 125–31, https://doi.org/10.1016/S0889-1591(02)00079-X.

125 **As a young doctor training:** Richard P. Richardson et al., "Peripheral Blood Leukocyte Kinetics Following in Vivo Lipopolysaccharide (LPS) Administration to Normal Human Subjects: Influence of Elicited Hormones and Cytokines," *Annals of Surgery* 210, no. 2 (August 1989): 239–45, https://journals.lww.com/annalsofsurgery/abstract/1989/08000/peripheral_blood_leukocyte_kinetics_following_in.18.aspx.

126 **known as *cytokine storm*:** David C. Fajgenbaum and Carl H. June, "Cytokine Storm," *New England Journal of Medicine* 383, no. 23 (December 2, 2020): 2255–73, https://doi.org/10.1056/NEJMra2026131.

126 **Later, in the lab:** Richard P. Richardson et al., "Peripheral Blood Leukocyte Kinetics."

126 **Many of these patients, prior:** Marc Udina et al., "Interferon-Induced Depression in Chronic Hepatitis C: A Systematic Review and Meta-Analysis," *Journal of Clinical Psychiatry* 73, no. 8 (August 15, 2012): 1128, https://doi.org/10.4088/JCP.12r07694.

127 **when we experience chronic stress:** Anna L. Marsland et al., "The Effects of Acute Psychological Stress on Circulating and Stimulated Inflammatory Markers: A Systematic Review and Meta-Analysis," *Brain, Behavior, and Immunity* 64 (January 12, 2017): 208–19, https://doi.org/10.1016/j.bbi.2017.01.011.

127 **risk developing depression:** Robert Dantzer et al., "From Inflammation to Sickness and Depression: When the Immune System Subjugates the Brain," *Nature Reviews Neuroscience* 9, no. 1 (January 2008): 46–56, https://doi.org/10.1038/nrn2297.

127 **Clinical studies support:** Christopher R. Pryce and Adriano Fontana, "Depression in Autoimmune Diseases," *Inflammation-Associated Depression: Evidence, Mechanisms and Implications* (May 25, 2016): 139–54, https://doi.org/10.1007/7854_2016_7; Jayaprakash Russell Ravan et al., "Autoimmune Rheumatic Diseases Masquerading as Psychiatric Disorders: A Case Series," *Mediterranean Journal of Rheumatology* 32, no. 2 (2021): 164–67, https://doi.org/10.31138/mjr.32.2.164.

128 **It's a very good question:** Il-Bin Kim, Jae-Hon Lee, and Seon-Cheol Park, "The Relationship Between Stress, Inflammation, and Depression," *Biomedicines* 10, no. 8 (August 9, 2022): 1929, https://doi.org/10.3390/biomedicines10081929.

128 **a predominant theory of depression:** Alec Coppen, "The Biochemistry of Affective Disorders," *The British Journal of Psychiatry* 113, no. 504 (November 1967): 1237–64, https://doi.org/10.1192/bjp.113.504.1237; Joanna Moncrieff et al., "The Serotonin Theory of Depression: A Systematic Umbrella Review

of the Evidence," *Molecular Psychiatry* 28, no. 8 (July 20, 2023): 3243–56, https://doi.org/10.1038/s41380-022-01661-0.

129 **A comprehensive review:** Moncrieff et al., "The Serotonin Theory."

129 **Genetic studies on the SERT:** Moncrieff et al., "The Serotonin Theory."

130 **administering SSRIs to animals:** Lucile Capuron et al., "Treatment of Cytokine-Induced Depression," *Brain, Behavior, and Immunity* 16, no. 5 (October 2002): 575–80, https://doi.org/10.1016/S0889-1591(02)00007-7; Nadine Herr, Christoph Bode, and Daniel Duerschmied, "The Effects of Serotonin in Immune Cells," *Frontiers in Cardiovascular Medicine* 4 (July 19, 2017): 48, https://doi.org/10.3389/fcvm.2017.00048.

130 **This is why SSRIs are:** Lina Wang et al., "Effects of SSRIs on Peripheral Inflammatory Markers in Patients with Major Depressive Disorder: A Systematic Review and Meta-Analysis," *Brain, Behavior, and Immunity* 79 (July 2019): 24–38, https://doi.org/10.1016/j.bbi.2019.02.021; Aric A. Prather et al., "Cytokine-Induced Depression During IFN-α Treatment: The Role of IL-6 and Sleep Quality," *Brain, Behavior, and Immunity* 23, no. 8 (November 2009): 1109–16, https://doi.org/10.1016/j.bbi.2009.07.001.

131 **We know this from a series:** Linda R. Watkins et al., "Blockade of Interleukin-1 Induced Hyperthermia by Subdiaphragmatic Vagotomy: Evidence for Vagal Mediation of Immune-Brain Communication," *Neuroscience Letters* 183, no. 1–2 (January 1995): 27–31, https://doi.org/10.1016/0304-3940(94)11105-R.

132 **When this barrier is breached:** Michael Camilleri, "Leaky Gut: Mechanisms, Measurement and Clinical Implications in Humans," *Gut* 68, no. 8 (August 2019): 1516–26, http://orcid.org/0000-0001-6472-7514.

133 **Other studies are exploring:** Erin A. Yamamoto and Trine N. Jørgensen, "Relationships Between Vitamin D, Gut Microbiome, and Systemic Autoimmunity," *Frontiers in Immunology* 10 (January 20, 2020): 499337, https://doi.org/10.3389/fimmu.2019.03141; Agata Twardowska et al., "Preventing Bacterial Translocation in Patients with Leaky Gut Syndrome: Nutrition and Pharmacological Treatment Options," *International Journal of Molecular Sciences* 23, no. 6 (2022): 3204, https://doi.org/10.3390/ijms23063204.

133 **Several labs, including mine and:** Bruno Bonaz, "Anti-inflammatory Effects of Vagal Nerve Stimulation with a Special Attention to Intestinal Barrier Dysfunction," *Neurogastroenterology & Motility* 34, no. 10 (September 12, 2022): e14456, https://doi.org/10.1111/nmo.14456; Harold A. Silverman et al., "Transient Receptor Potential Ankyrin-1-Expressing Vagus Nerve Fibers Mediate IL-1β Induced Hypothermia and Reflex Anti-inflammatory Responses," *Molecular Medicine* 29, no. 1 (January 18, 2023): 4, https://doi.org/10.1186/s10020-022-00590-6.

133 **details in the endnotes:** The final answers are not yet in, but areas under active consideration or ongoing study for how vagus nerve dysfunction or impairment can lead to depression include impaired signaling in the dorsal motor nucleus of the vagus (DMNV), which exerts crucial parasympathetic control over visceral organs, and can affect gastrointestinal and cardiac functions disrupted in depression. Dysfunction in the nucleus tractus solitarius (NTS), the primary hub in the brain stem for information flowing

up the vagus nerve, can impact the processing of visceral information and emotional responses, contributing to depressive symptoms.

Projections from the NTS to the amygdala, that brain region involved in processing emotions, may be impaired in depression, and this altered connectivity between the NTS and the amygdala can lead to heightened emotional reactivity and dysregulation. Other vagus nerve projections to the prefrontal cortex may affect the cognitive aspects of depression, such as decision-making and attention. And altered signaling in the parabrachial nucleus, which normally relays vagus nerve signals to the limbic system, can contribute to the emotional and autonomic symptoms of depression.

134 **one pivotal trial:** As measured by standard scales like the Hamilton Depression Rating Scale; T. E. Schlaepfer et al., "Vagus Nerve Stimulation for Depression: Efficacy and Safety in a European Study," *Psychological Medicine* 38, no. 5 (January 4, 2008): 651–61, https://doi.org/10.1017/S0033291707001924.

134 **A much larger clinical trial:** Called "Prospective, Multi-center, Randomized Controlled Blinded Trial Demonstrating the Safety and Effectiveness of VNS Therapy® System as Adjunctive Therapy Versus a No Stimulation Control in Subjects with Treatment-Resistant Depression (RECOVER)," last updated March 13, 2024, https://clinicaltrials.gov/study/NCT03887715.

7. Outside-In Stimulation to Regulate Body Weight, Treat Diabetes, and More

139 **obese individuals worldwide:** World Health Organization, "Obesity and Overweight," March 1, 2024, https://www.who.int/news-room/fact-sheets/detail/obesity-and-overweight.

140 **An expanding field of obesity:** Sandra A. Fryhofer, MD, Chair, presenter, "Report of the Council on Science and Public Health: Is Obesity a Disease?," American Medical Association, 2013, https://www.ama-assn.org/sites/ama-assn.org/files/corp/media-browser/public/about-ama/councils/Council%20Reports/council-on-science-public-health/a13csaph3.pdf.

140 **Millions of people today:** D. K. Arulmozhi and B. Portha, "GLP-1 Based Therapy for Type 2 Diabetes," *European Journal of Pharmaceutical Sciences* 28, no. 1–2 (May 2006): 96–108, https://doi.org/10.1016/j.ejps.2006.01.003.

140 **This useful side effect:** Ryan Hogg, "Brands like Walmart and Nestle Are Right to Worry About Wegovy. Novo Nordisk Sold $900m of the Weight Loss Drug Last Quarter—and 95% Was in the U.S.," *Fortune*, November 2, 2023, https://fortune.com/europe/2023/11/02/novo-nordisk-walmart-nestleweight-loss-drug-us/.

140 **GLP-1 agonists stimulate the vagus:** Astrid Plamboeck et al., "The Effect of Exogenous GLP-1 on Food Intake Is Lost in Male Truncally Vagotomized Subjects with Pyloroplasty," *American Journal of Physiology-Gastrointestinal and Liver Physiology* 304, no. 12 (June 15, 2013): G1117–27, https://doi.org/10.1152/ajpgi.00035.2013.

141 **We established a collaboration:** Founded by Thomas Edison, this spectac-

ular scenic campus sits high on a bluff overlooking the Mohawk River in Niskayuna, NY (https://www.ge.com/research/research-engine/rd-facilities/niskayuna). I treasure a photo from one visit, taken as I stood in the lobby next to Thomas Edison's wooden desk.

141 **A "Western diet":** Anette Christ, Mario Lauterbach, and Eicke Latz, "Western Diet and the Immune System: An Inflammatory Connection," *Immunity* 51, no. 5 (November 2019): 794–811, https://doi.org/10.1016/j.immuni.2019.09.020.

142 **worked with our team:** Especially the brilliant and dedicated graduate student Tomás Huerta.

142 **We divided the obese mice:** Tomás S. Huerta et al., "Targeted Peripheral Focused Ultrasound Stimulation Attenuates Obesity-Induced Metabolic and Inflammatory Dysfunctions," *Scientific Reports* 11, no. 1 (March 3, 2021): 5083, https://doi.org/10.1038/s41598-021-84330-6.

142 **"these findings suggest a previously":** Huerta et al., "Targeted Peripheral Focused Ultrasound."

144 **Peripheral focused ultrasound (pFUS):** William J. Tyler et al., "Remote Excitation of Neuronal Circuits Using Low-Intensity, Low-Frequency Ultrasound," *PLOS One* 3, no. 10 (October 29, 2008), https://doi.org/10.1371/journal.pone.0003511: e3511; Victoria Cotero et al., "Peripheral Focused Ultrasound Neuromodulation (pFUS)," *Journal of Neuroscience Methods* 341 (July 2020): 108721, https://doi.org/10.1016/j.jneumeth.2020.108721.

144 **clinical trials place focused ultrasound:** Shi-Chun Bao et al., "Peripheral Focused Ultrasound Stimulation and Its Applications: From Therapeutics to Human-Computer Interaction," *Frontiers in Neuroscience* 17 (April 13, 2023): 1115946, https://doi.org/10.3389/fnins.2023.1115946; Zahra Izadifar et al., "An Introduction to High Intensity Focused Ultrasound: Systematic Review on Principles, Devices, and Clinical Applications," *Journal of Clinical Medicine* 9, no. 2 (2020): 460, https://doi.org/10.3390/jcm9020460.

144 **by a different mechanism:** Hongchae Baek, Ki Joo Pahk, and Hyungmin Kim, "A Review of Low-Intensity Focused Ultrasound for Neuromodulation," *Biomedical Engineering Letters* 7 (January 9, 2017): 135–42, https://doi.org/10.1007/s13534-016-0007-y.

145 **When food enters the stomach:** Hailley Loper et al., "Both High Fat and High Carbohydrate Diets Impair Vagus Nerve Signaling of Satiety," *Scientific Reports* 11, no. 1 (May 17, 2021): 10394, https://doi.org/10.1038/s41598-021-89465-0.

145 **In a recent study, thin mice:** Peter J. Turnbaugh et al., "An Obesity-Associated Gut Microbiome with Increased Capacity for Energy Harvest," *Nature* 444, no. 7122 (December 21, 2006): 1027–31, https://doi.org/10.1038/nature05414.

145 **And here's the proof:** Yunpeng Liu and Paul Forsythe, "Vagotomy and Insights into the Microbiota-Gut-Brain Axis," *Neuroscience Research* 168 (July 2021): 20–27, https://doi.org/10.1016/j.neures.2021.04.001; Edward S. Bliss and Eliza Whiteside, "The Gut-Brain Axis, the Human Gut Microbiota and Their Integration in the Development of Obesity," *Frontiers in Physiology* 9

(July 11, 2018): 368061, https://doi.org/10.3389/fphys.2018.00900; Marialetizia Rastelli, Claude Knauf, and Patrice D. Cani, "Gut Microbes and Health: A Focus on the Mechanisms Linking Microbes, Obesity, and Related Disorders," *Obesity* 26, no. 5 (May 2018): 792–800, https://doi.org/10.1002/oby.22175.

146   **They noticed that blocking:** Daniel I. Brierley and Guillaume de Lartigue, "Reappraising the Role of the Vagus Nerve in GLP-1-Mediated Regulation of Eating," *British Journal of Pharmacology* 179, no. 4 (February 2022): 584–99, https://doi.org/10.1111/bph.15603.

146   **In other mouse experiments:** Susanna Longo, Stefano Rizza, and Massimo Federici, "Microbiota-Gut-Brain Axis: Relationships Among the Vagus Nerve, Gut Microbiota, Obesity, and Diabetes," *Acta Diabetologica* 60, no. 8 (2023): 1007–17; Jean-Philippe Krieger, Wolfgang Langhans, and Shin J. Lee, "Vagal Mediation of GLP-1's Effects on Food Intake and Glycemia," *Physiology & Behavior* 152 (2015): 372–80.

147   **not in the vagotomized subjects:** Astrid Plamboeck et al., "The Effect of Exogenous GLP-1 on Food Intake Is Lost in Male Truncally Vagotomized Subjects with Pyloroplasty," *American Journal of Physiology Gastrointestinal and Liver Physiology* 15, no. 304 (June 15, 2013): G1117–27, https://doi.org/10.1152/ajpgi.00035.2013.

147   **GE HealthCare has announced:** GE HealthCare, "GE HealthCare and Novo Nordisk to Collaborate to Advance Novel Non-Invasive Treatment for Type 2 Diabetes and Obesity with Ultrasound," press release, October 19, 2023, https://www.gehealthcare.com/about/newsroom/press-releases/ge-healthcare-and-novo-nordisk-to-collaborate-to-advance-novel-non-invasive-treatment-for-type-2-diabetes-and-obesity-with-ultrasound?npclid=botnpclid.

148   **Our colleagues in Minnesota:** Daniel P. Zachs et al., "Noninvasive Ultrasound Stimulation of the Spleen to Treat Inflammatory Arthritis," *Nature Communications* 10, no. 1 (March 12, 2019): 951, https://doi.org/10.1038/s41467-019-08721-0.

148   **Focused ultrasound activated the splenic:** Victoria Cotero et al., "Noninvasive Sub-Organ Ultrasound Stimulation for Targeted Neuromodulation," *Nature Communications* 10, no. 1 (March 2019): 952, https://doi.org/10.1038/s41467-019-08750-9.

148   **Furthermore, focused ultrasound stimulation:** Mauricio Rosas-Ballina et al., "Acetylcholine-Synthesizing T Cells Relay Neural Signals in a Vagus Nerve Circuit," *Science* 334, no. 6052 (2011): 98–101, https://doi.org/10.1126/science.1209985; Mauricio Rosas-Ballina et al., "Splenic Nerve Is Required for Cholinergic Anti-inflammatory Pathway Control of TNF in Endotoxemia," *Proceedings of the National Academy of Sciences* 105, no. 31 (August 5, 2008): 11008–13, https://doi.org/10.1073/pnas.0803237105.

149   **Based on these mice studies:** Sangeeta S. Chavan and Kevin J. Tracey, "Essential Neuroscience in Immunology," *The Journal of Immunology* 198, no. 9 (May 1, 2017): 3389–97, https://doi.org/10.4049/jimmunol.1601613.

149   **The focused ultrasound power:** Stavros Zanos et al., "Focused Ultrasound Neuromodulation of the Spleen Activates an Anti-Inflammatory Response

149 **Using a laboratory method:** Frieda A. Koopman et al., "Vagus Nerve Stimulation Inhibits Cytokine Production and Attenuates Disease Severity in Rheumatoid Arthritis," *Proceedings of the National Academy of Sciences* 113, no. 29 (July 5, 2016): 8284–89, https://doi.org/10.1073/pnas.1605635113.

150 **Our colleague Stavros Zanos:** "Reconstructing Vagal Anatomy," Project Details, NIH RePORTER, https://reporter.nih.gov/project-details/10723189.

151 **Guided by the map:** Stephan L. Blanz et al., "Spatially Selective Stimulation of the Pig Vagus Nerve to Modulate Target Effect Versus Side Effect," *Journal of Neural Engineering* 20, no. 1 (February 22, 2023): 016051, https://doi.org/10.1088/1741-2552/acb3fd.

152 **Patient compliance with self-administered:** Fernandez-Lazaro et al., "Adherence to Treatment and Related Factors Among Patients with Chronic Conditions in Primary Care: A Cross-Sectional Study," *BMC Family Practice* 20, no. 132 (September 4, 2019): 1–12, https://doi.org/10.1186/s12875-019-1019-3.

8. The Ear-Brain-Body Connection: Over-the-Counter Devices with Many Potential Benefits

155 **epigraph:** "Maya Angelou's 'Note to Self,'" CBS News, September 4, 2013, https://www.cbsnews.com/news/maya-angelous-note-to-self/.

157 **his own clinical report:** Ulf Andersson, Gustaf Kranck, and Dr. Fintan Nagle, "Case Study—Findings About the Vagus Test," published as part of Vagus Health Ltd. scientific findings, Stockholm, Sweden, December 24, 2022, https://vagus.co/case-study-findings-about-the-vagus-test/.

159 *Arnold nerve reflex*: Mohsin F. Butt et al., "The Anatomical Basis for Transcutaneous Auricular Vagus Nerve Stimulation," *Journal of Anatomy* 236, no. 4 (November 19, 2020): 588–611, https://doi.org/10.1111/joa.13122.

159 **Acupuncture of the ear:** Andreas Wirz-Ridolfi, "The History of Ear Acupuncture and Ear Cartography: Why Precise Mapping of Auricular Points Is Important," *Medical Acupuncture* 31, no. 3 (June 17, 2019): 145–56, https://doi.org/10.1089/acu.2019.1349.

160 **Dr. Nogier launched the new field:** Raphaël Nogier, "How Did Paul Nogier Establish the Map of the Ear?," *Medical Acupuncture* 26, no. 2 (April 2014): 76–83, https://doi.org/10.1089/acu.2014.1035.

161 **Later publication of his:** X. L. Ye, "The New Discovery of Acupuncture Abroad—The Introduction to Auricular Acupuncture Therapy," *Shanghai Journal of Traditional Chinese Medicine* 12 (1958): 45–8.

161 **In 2001, one of the earliest:** A. V. Zamotrinsky, B. Kondratiev, and Jan Willem de Jong, "Vagal Neurostimulation in Patients with Coronary Artery Disease," *Autonomic Neuroscience* 88, no. 1–2 (April 12, 2001): 109–16, https://doi.org/10.1016/S1566-0702(01)00227-2.

161 **The researchers observed significant decreases:** Ronald G. Garcia et al., "Optimization of Respiratory-Gated Auricular Vagus Afferent Nerve

Stimulation for the Modulation of Blood Pressure in Hypertension," *Frontiers in Neuroscience* 16 (December 8, 2022): 1038339, https://doi.org/10.3389/fnins.2022.1038339.

161 **There are piles of clinical:** Angela Yun Kim et al., "Safety of Transcutaneous Auricular Vagus Nerve Stimulation (taVNS): A Systematic Review and Meta-Analysis," *Scientific Reports* 12, no. 1 (December 2022): 22055, https://doi.org/10.1038/s41598-022-25864-1; Yusuf Ozgur Cakmak, "Concerning Auricular Vagal Nerve Stimulation: Occult Neural Networks," *Frontiers in Human Neuroscience* 13 (2019): 421, https://doi.org/10.3389/fnhum.2019.00421; Chaoren Tan et al., "The Efficacy and Safety of Transcutaneous Auricular Vagus Nerve Stimulation in the Treatment of Depressive Disorder: A Systematic Review and Meta-Analysis of Randomized Controlled Trials," *Journal of Affective Disorders* (September 15, 2023): 37–49, https://doi.org/10.1016/j.jad.2023.05.048.

162 **A physician in ancient Rome:** Francis Sahngun Nahm, "From the Torpedo Fish to the Spinal Cord Stimulator," *The Korean Journal of Pain* 33, no. 2 (April 1, 2020): 97–98, https://doi.org/10.3344/kjp.2020.33.2.97.

162 **"gate" in the spinal cord:** Ronald Melzack and Patrick D. Wall, "Pain Mechanisms: A New Theory: A Gate Control System Modulates Sensory Input from the Skin Before It Evokes Pain Perception and Response," *Science* 150, no. 3699 (November 19, 1965): 971–79, https://doi.org/10.1126/science.150.3699.971.

163 **The effects of various types:** Ashraf N. H. Gerges et al., "Clinical Application of Transcutaneous Auricular Vagus Nerve Stimulation: A Scoping Review," *Disability and Rehabilitation* (February 16, 2024): 1–31, https://doi.org/10.1080/09638288.2024.2313123.

164 **the innervation of ears:** Cakmak, "Concerning Auricular Vagal Nerve Stimulation."

165 *auricular electrical stimulation,* **or AES:** Cakmak, "Concerning Auricular Vagal Nerve Stimulation."

167 **Four days later, we repeated:** Raphaël Nogier, "History of Auriculotherapy: Additional Information and New Developments," *Medical Acupuncture* 33, no. 6 (December 16, 2021): 410–19, https://doi.org/10.1089/acu.2021.0075.

168 **In a controlled study using:** Eleni Frangos, Jens Ellrich, and Barry R. Komisaruk, "Non-Invasive Access to the Vagus Nerve Central Projections via Electrical Stimulation of the External Ear: fMRI Evidence in Humans," *Brain Stimulation* 8, no. 3 (May–June 2015): 624–36, https://doi.org/10.1016/j.brs.2014.11.018; Eleni Frangos, "Non-Invasive Access to the Vagus Nerve and Its Projections in Humans: fMRI Evidence" (PhD dissertation, Rutgers University-Graduate School-Newark, 2014); Natalia Yakunina, Sam Soo Kim, and Eui-Cheol Nam, "Optimization of Transcutaneous Vagus Nerve Stimulation Using Functional MRI," *Neuromodulation: Technology at the Neural Interface* 20, no. 3 (April 2017): 290–300, https://doi.org/10.1111/ner.12541.

168 **electrodes placed on the scalp:** A. J. Fallgatter et al., "Far Field Potentials from the Brainstem After Transcutaneous Vagus Nerve Stimulation," *Jour-

nal of Neural Transmission 110 (January 1, 2003): 1437–43, https://doi.org /10.1007/s00702-003-0087-6.

168 **Analysis of these brain stem:** Thomas Polak et al., "Far Field Potentials from Brainstem After Transcutaneous Vagus Nerve Stimulation: Optimization of Stimulation and Recording Parameters," *Journal of Neural Transmission* 116 (September 1, 2009): 1237–42, https://doi.org/10.1007/s00702-009-0282-1.

168 **Healthy volunteers receiving:** Omer Sharon, Firas Fahoum, and Yuval Nir, "Transcutaneous Vagus Nerve Stimulation in Humans Induces Pupil Dilation and Attenuates Alpha Oscillations," *Journal of Neuroscience* 41, no. 2 (January 13, 2021): 320–30, https://doi.org/10.1523/JNEUROSCI.1361-20.2020.

168 **This enhancement has been attributed:** Rico Fischer et al., "Transcutaneous Vagus Nerve Stimulation (tVNS) Enhances Conflict-Triggered Adjustment of Cognitive Control," *Cognitive, Affective, & Behavioral Neuroscience* 18 (August 2018): 680–93, https://doi.org/10.3758/s13415-018-0596-2; Jelle W. R. Van Leusden, Roberta Sellaro, and Lorenza S. Colzato, "Transcutaneous Vagal Nerve Stimulation (tVNS): A New Neuromodulation Tool in Healthy Humans?," *Frontiers in Psychology* 6 (February 10, 2015): 127729, https://doi .org/10.3389/fpsyg.2015.00102; Vesna M. Van Midden et al., "The Effects of Transcutaneous Auricular Vagal Nerve Stimulation on Cortical GABAergic and Cholinergic Circuits: A Transcranial Magnetic Stimulation Study," *European Journal of Neuroscience* 57, no. 12 (June 2023): 2160–73, https://doi.org/10.1111/ejn.16004.

169 **oxygen in the prefrontal cortex:** Liyan Peng et al., "Transauricular Vagus Nerve Stimulation at Auricular Acupoints Kidney (CO10), Yidan (CO11), Liver (CO12) and Shenmen (TF4) Can Induce Auditory and Limbic Cortices Activation Measured by fMRI," *Hearing Research* 359 (March 2018): 1–12, https://doi.org/10.1016/j.heares.2017.12.003.

169 **ear TENS in epilepsy patients:** Peijing Rong et al., "Transcutaneous Vagus Nerve Stimulation for Refractory Epilepsy: A Randomized Controlled Trial," *Clinical Science* (April 1, 2014): CS20130518, https://doi.org/10.1042 /CS20130518; Sebastian Bauer et al., "Transcutaneous Vagus Nerve Stimulation (tVNS) for Treatment of Drug-Resistant Epilepsy: A Randomized, Double-Blind Clinical Trial (cMPsE02)," *Brain Stimulation* 9, no. 3 (January 2016): 356–63, https://doi.org/10.1016/j.brs.2015.11.003; Marios Lampros et al., "Transcutaneous Vagus Nerve Stimulation (t-VNS) and Epilepsy: A Systematic Review of the Literature," *Seizure* 91 (October 2021): 40–48; https:// doi.org/10.1016/j.seizure.2021.05.017.

170 **inhibition of cytokine production:** Aisling Tynan, Michael Brines, and Sangeeta S. Chavan, "Control of Inflammation Using Non-Invasive Neuromodulation: Past, Present and Promise," *International Immunology* 34, no. 2 (February 2022): 119–28, https://doi.org/10.1093/intimm/dxab073.

170 **In a series of peer:** Meghan E. Addorisio et al., "Investigational Treatment of Rheumatoid Arthritis with a Vibrotactile Device Applied to the External Ear," *Bioelectronic Medicine* 5 (April 17, 2019): 1–11, https://doi.org/10.1186 /s42234-019-0020-4; Cynthia Aranow et al., "Transcutaneous Auricular Vagus Nerve Stimulation Reduces Pain and Fatigue in Patients with Systemic

Lupus Erythematosus: A Randomised, Double-Blind, Sham-Controlled Pilot Trial," *Annals of the Rheumatic Diseases* 80, no. 2 (February 2021): 203–08, https://doi.org/10.1136/annrheumdis-2020-217872; Kumail Merchant et al., "Transcutaneous Auricular Vagus Nerve Stimulation (taVNS) for the Treatment of Pediatric Nephrotic Syndrome: A Pilot Study," *Bioelectronic Medicine* 8, no. 1 (January 26, 2022): 1, https://doi.org/10.1186/s42234-021-00084-6; Benjamin Sahn et al., "Transcutaneous Auricular Vagus Nerve Stimulation Attenuates Inflammatory Bowel Disease in Children: A Proof-of-Concept Clinical Trial," *Bioelectronic Medicine* 9, no. 1 (October 18, 2023): 23, https://doi.org/10.1186/s42234-023-00124-3.

170 **The FDA has approved the commercial:** Imran S. Qureshi et al., "Auricular Neural Stimulation as a New Non-Invasive Treatment for Opioid Detoxification," *Bioelectronic Medicine* 6, no. 1 (March 30, 2020): 7, https://doi.org/10.1186/s42234-020-00044-6.

171 **cluster and migraine headache:** Jonathan Y. Y. Yap et al., "Critical Review of Transcutaneous Vagus Nerve Stimulation: Challenges for Translation to Clinical Practice," *Frontiers in Neuroscience* 14 (April 27, 2020): 495370, https://doi.org/10.3389/fnins.2020.00284.

171 **A 2022 study:** Duyan Geng et al., "The Effect of Transcutaneous Auricular Vagus Nerve Stimulation on HRV in Healthy Young People," *PLOS One* 17, no. 2 (February 10, 2022): e0263833, https://doi.org/10.1371/journal.pone.0263833.

172 **In one study, performed:** Peijing Rong et al., "Effect of Transcutaneous Auricular Vagus Nerve Stimulation on Major Depressive Disorder: A Nonrandomized Controlled Pilot Study," *Journal of Affective Disorders* 195 (May 2016): 172–79, https://doi.org/10.1016/j.jad.2016.02.031.

172 **A review of twelve clinical:** Tan et al., "The Efficacy and Safety of Transcutaneous Auricular Vagus Nerve Stimulation in the Treatment of Depressive Disorder: A Systematic Review and Meta-Analysis of Randomized Controlled Trials," *Journal of Affective Disorders* 337 (September 15, 2023): 37–49, https://doi.org/10.1016/j.jad.2023.05.048.

174 **Until now, most clinical trials:** Yap et al., "Critical Review."

174 **Currently, we do not know:** Butt et al., "The Anatomical Basis."

174 **And since nerve locations:** Cakmak, "Concerning Auricular Vagal Nerve Stimulation."

174 **One study of insomnia patients:** Shuai Zhang et al., "Effects of Transcutaneous Auricular Vagus Nerve Stimulation on Brain Functional Connectivity of Medial Prefrontal Cortex in Patients with Primary Insomnia," *The Anatomical Record* 304, no. 11 (2021): 2426–35, https://doi.org/10.1002/ar.24785.

174 **Sangeeta and I also reported:** Addorisio et al., "Investigational Treatment of Rheumatoid Arthritis."

Part Three: Great Expectations

177 **art:** Valentin A. Pavlov and Kevin J. Tracey, "The Vagus Nerve and the Inflammatory Reflex—Linking Immunity and Metabolism," *Nature Reviews*

Endocrinology 8, no. 12 (December 2012): 743–54, https://doi.org/10.1038/nrendo.2012.189.

9. Meditation and Breathwork

179 **epigraph:** Theodore Sturgeon, writer, *Star Trek: The Original Series*, "Amok Time," season 2, episode 1, 1968.

181 **several bestsellers on the interplay:** E.g., His Holiness the Dalai Lama and Howard C. Cutler, *The Art of Happiness: A Handbook for Living*, 10th anniversary ed. (New York: Riverhead Books, 1998); His Holiness the Dalai Lama and Archbishop Desmond Tutu with Douglas Abrams, *The Book of Joy: Lasting Happiness in a Changing World* (New York: Avery, 2016).

182 **Elizabeth's discovery of telomeres:** Elizabeth Blackburn and Elissa Epel, *The Telomere Effect: A Revolutionary Approach to Living Younger, Healthier, Longer* (New York: Grand Central Publishing, 2017). Telomeres are the molecular structures located on the ends of your DNA molecules, sometimes likened to the plastic thingies wrapped around the ends of your shoelaces, which almost nobody remembers are called *aglets*. The aglets of vitality, telomeres keep your chromosomes from fraying and slow or minimize the effects of aging.

183 **For example, one systematic review:** Kieran C. R. Fox et al., "Functional Neuroanatomy of Meditation: A Review and Meta-Analysis of 78 Functional Neuroimaging Investigations," *Neuroscience & Biobehavioral Reviews* 65 (June 2016): 208–28, https://doi.org/10.1016/j.neubiorev.2016.03.021.

185 **A review of nineteen randomized:** Lydia Brown et al., "The Effects of Mindfulness and Meditation on Vagally Mediated Heart Rate Variability: A Meta-Analysis," *Psychosomatic Medicine* 83, no. 6 (July–August 2021): 631–40, https://doi.org/10.1097/PSY.0000000000000900.

185 **Given the public health implications:** Kenneth Jamerson et al., "Meditation and Cardiovascular Risk Reduction," *Journal of the American Heart Association* 6 (September 28, 2017): e002218, https://doi.org/10.1161/JAHA.117.002218.

185 **A study that reviewed forty-eight:** Thomas J. Dunn and Mirena Dimolareva, "The Effect of Mindfulness-Based Interventions on Immunity-Related Biomarkers: A Comprehensive Meta-Analysis of Randomised Controlled Trials," *Clinical Psychology Review* 92 (March 2022): 102124, https://doi.org/10.1016/j.cpr.2022.102124.

186 **psychological benefits of meditation:** Roderik J. S. Gerritsen and Guido P. H. Band, "Breath of Life: the Respiratory Vagal Stimulation Model of Contemplative Activity," *Frontiers in Human Neuroscience* 12 (October 8, 2018): 393151, https://doi.org/10.3389/fnhum.2018.00397.

187 **a fascinating TED Talk:** Andy Puddicombe, "All It Takes Is 10 Mindful Minutes," TED Talk, TEDSalon London, 9 min., 30 sec., November 2012, https://www.ted.com/talks/andy_puddicombe_all_it_takes_is_10_mindful_minutes?language=en.

188 **"receive total consciousness":** "*Caddyshack* (1980): Bill Murray: Carl

NOTES

Spackler," Quotes, IMDb, https://www.imdb.com/title/tt0080487/characters/nm0000195.

190 **One study from Harvard:** Xiao Ma et al., "The Effect of Diaphragmatic Breathing on Attention, Negative Affect and Stress in Healthy Adults," *Frontiers in Psychology* 8 (June 5, 2017): 234806, https://doi.org/10.3389/fpsyg.2017.00874.

191 **In short, diaphragmatic breathing:** For a good overall discussion of this topic, I recommend James Nestor's *Breath: The New Science of a Lost Art* (New York: Riverhead Books, 2020).

191 ***Like many, probably, I was:*** Joe Rogan, host, *The Joe Rogan Experience,* podcast, episode 712, "Wim Hof Method," October 20, 2015, https://www.jrepodcast.com/episode/joe-rogan-experience-712-wim-hof/#google_vignette.

192 **"body and the brain together":** Joe Rogan, host, *The Joe Rogan Experience.*

192 **"YOU ARE STRONGER":** *Wim Hof Method,* website home page, accessed March 18, 2024, https://www.wimhofmethod.com.

192 **"full capacity of [his] physiology":** Joe Rogan, host, *The Joe Rogan Experience.*

192 **influence its functioning at will:** Joe Rogan, host, *The Joe Rogan Experience.*

193 **thirty-breath round:** "Breathing Exercises," Wim Hof Method, accessed March 18, 2024, https://www.wimhofmethod.com/breathing-exercises.

195 **The paper summarized the results:** Matthijs Kox et al., "Voluntary Activation of the Sympathetic Nervous System and Attenuation of the Innate Immune Response in Humans," *Proceedings of the National Academy of Sciences* 111, no. 20 (May 5, 2014): 7379–84, https://doi.org/10.1073/pnas.1322174111.

195 **The study's lead author:** Kox et al., "Voluntary Activation."

197 **Surprisingly, the simulated:** Tom van der Poll et al., "Epinephrine Inhibits Tumor Necrosis Factor-Alpha and Potentiates Interleukin 10 Production During Human Endotoxemia," *The Journal of Clinical Investigation* 97, no. 3 (February 1, 1996): 713–19, https://doi.org/10.1172/JCI118469.

197 **negative feedback mechanism:** Tom Van der Poll et al., "Noradrenaline Inhibits Lipopolysaccharide-Induced Tumor Necrosis Factor and Interleukin 6 Production in Human Whole Blood," *Infection and Immunity* 62, no. 5 (May 1, 1994): 2046–50, https://doi.org/0.1128/iai.62.5.2046-2050.1994. PMID: 8168970.

197 **Recall the sheep:** Julia Shanks et al., "Cardiac Vagal Nerve Activity Increases During Exercise to Enhance Coronary Blood Flow," *Circulation Research* 133, no. 7 (August 29, 2023): 559–71, https://doi.org/10.1161/CIRCRESAHA.123.323017.

198 **research has shown that slow**: Chen Hsu Wang, Hui-Wen Yang, Han-Luen Huang, Cheng-Yi Hsiao, Bun-Kai Jiu, Chen Lin, and Men-Tzung Lo, "Long-Term Effect of Device-Guided Slow Breathing on Blood Pressure Regulation and Chronic Inflammation in Patients with Essential Hypertension Using a Wearable ECG Device," *Acta Cardiologica Sinica* 37, no. 2 (2021): 195, https://doi.org/10.6515/ACS.202103_ 37(2).20200907A.

198 **As the Harvard-Beijing study:** Caroline Sevoz-Couche and Sylvain Laborde,

"Heart Rate Variability and Slow-Paced Breathing: When Coherence Meets Resonance," *Neuroscience & Biobehavioral Reviews* 135 (April 2022): 104576, https://doi.org/10.1016/j.neubiorev.2022.104576.

199 **Each of these breathwork exercises:** Melis Yilmaz Balban et al., "Brief Structured Respiration Practices Enhance Mood and Reduce Physiological Arousal," *Cell Reports Medicine* 4, no. 1 (January 17, 2023), https://doi.org/10.1016/j.xcrm.2022.100895.

199 **This protocol resulted in a:** Wang et al., "Long-Term Effect of Device-Guided Slow Breathing."

200 **Within two weeks, patients in:** Elisabeth Maria Balint et al., "A Randomized Clinical Trial to Stimulate the Cholinergic Anti-Inflammatory Pathway in Patients with Moderate COVID-19-Pneumonia Using a Slow-Paced Breathing Technique," *Frontiers in Immunology* 13 (October 3, 2022): 928979, https://doi.org/10.3389/fimmu.2022.928979.

200 **Out of the one hundred:** Per work from Stephen Liberles at Harvard University and from Yaakov Levine at SetPoint Medical.

202 **Breathwork is taught in some:** Tanya G. K. Bentley et al., "Slow-Breathing Curriculum for Stress Reduction in High School Students: Lessons Learned from a Feasibility Pilot," *Frontiers in Rehabilitation Sciences* 3 (July 2022): 864079, https://doi.org/10.3389/fresc.2022.864079.

202 **"It feels wonderful":** Joe Rogan, host, *The Joe Rogan Experience*.

203 **"Everybody is able to do it":** Joe Rogan, host, *The Joe Rogan Experience*.

10. Cold and Exercise

205 **epigraph:** Wim Hof (@Iceman_Hof), "The cold is a teacher. It's merciless," Twitter (now X), December 4, 2022, https://x.com/Iceman_Hof/status/1599418604956663811?lang=en.

210 **He treated malaria patients:** Robert Allan et al., "Cold for Centuries: A Brief History of Cryotherapies to Improve Health, Injury and Post-Exercise Recovery," *European Journal of Applied Physiology* 122, no. 5 (2022): 1153–62, https://doi.org/10.1007/s00421-022-04915-5.

211 **one study in the early 2000s:** Tiina M. Mäkinen et al., "Autonomic Nervous Function During Whole-Body Cold Exposure Before and After Cold Acclimation," *Aviation, Space, and Environmental Medicine* 79, no. 9 (September 2008): 875–82, https://doi.org/10.3357/asem.2235.2008.

211 **The lead researcher noted that:** Mäkinen et al., "Autonomic Nervous Function."

213 **Intercostal and pectoral muscles:** Otto Muzik, Kaice T. Reilly, and Vaibhav A. Diwadkar, "'Brain Over Body'—A Study on the Willful Regulation of Autonomic Function During Cold Exposure," *NeuroImage* 172 (May 15, 2018): 632–41, https://doi.org/10.1016/j.neuroimage.2018.01.067.

213 **"the magic began":** Joe Rogan, host, *The Joe Rogan Experience*.

214 **One French study:** Xavier Guillot et al., "Local Ice Cryotherapy Decreases Synovial Interleukin 6, Interleukin 1β, Vascular Endothelial Growth Fac-

tor, Prostaglandin-E2, and Nuclear Factor Kappa B p65 in Human Knee Arthritis: A Controlled Study," *Arthritis Research & Therapy* 21 (July 2019): 1–11, https://doi.org/10.1186/s13075-019-1965-0.

215 **Consider a study:** Milda Eimonte et al., "Residual Effects of Short-Term Whole-Body Cold-Water Immersion on the Cytokine Profile, White Blood Cell Count, and Blood Markers of Stress," *International Journal of Hyperthermia* 38, no. 1 (2021): 696–707, https://doi.org/10.1080/02656736.2021.1915504.

216 **"repeated cold water immersions":** L. Janský et al., "Immune System of Cold-Exposed and Cold-Adapted Humans," *European Journal of Applied Physiology and Occupational Physiology* 72 (1996): 445–50, https://doi.org/10.1007/BF00242274.

216 **This is a very different scenario:** Frieda A. Koopman et al., "Vagus Nerve Stimulation Inhibits Cytokine Production and Attenuates Disease Severity in Rheumatoid Arthritis," *Proceedings of the National Academy of Sciences* 113, no. 29 (July 5, 2016): 8284–89, https://doi.org/10.1073/pnas.1605635113.

217 **Because we observed significantly lower:** Koopman et al., "Vagus Nerve Stimulation."

217 **This small study:** Shawn G. Rhind et al., "Intracellular Monocyte and Serum Cytokine Expression Is Modulated by Exhausting Exercise and Cold Exposure," *American Journal of Physiology—Regulatory, Integrative and Comparative Physiology* 281, no. 1 (July 1, 2001): R66–75, https://doi.org/10.1152/ajpregu.2001.281.1.R66.

217 **This impairment could be due:** Koopman et al., "Vagus Nerve Stimulation."

221 **"If exercise could be packed":** Robert N. Butler, "Public Interest Report no. 23: Exercise, the Neglected Therapy," *The International Journal of Aging and Human Development* 8, no. 2 (March 1978): 193–95, https://doi.org/10.2190/AM1W-RABB-4PJY-P1PK.

222 **"exercise training increases parasympathetic tone":** Wayne C. Levy et al., "Effect of Endurance Exercise Training on Heart Rate Variability at Rest in Healthy Young and Older Men," *The American Journal of Cardiology* 82, no. 10 (November 15, 1998): 1236–41, https://doi.org/10.1016/s0002-9149(98)00611-0.

222 **Numerous large studies confirm:** Kyle Mandsager et al., "Association of Cardiorespiratory Fitness with Long-Term Mortality Among Adults Undergoing Exercise Treadmill Testing," *JAMA Network Open* 1, no. 6 (October 19, 2018): e183605, https://doi.org/10.1001/jamanetworkopen.2018.3605.

223 **popularized by Tony Cerami:** Ronald J. Koenig et al., "Correlation of Glucose Regulation and Hemoglobin A1c in Diabetes Mellitus," *New England Journal of Medicine* 295, no. 8 (August 19, 1976): 417–20, https://doi.org/10.1056/NEJM197608192950804.

224 **They also found:** Kyle L. Timmerman et al., "Exercise Training-Induced Lowering of Inflammatory (CD14+ CD16+) Monocytes: A Role in the Anti-Inflammatory Influence of Exercise?," *Journal of Leucocyte Biology* 84, no. 5 (July 29, 2008): 1271–78, https://doi.org/10.1189/jlb.0408244.

224 **another study with Richard Sloan:** Richard P. Sloan et al., "Aerobic Exercise Attenuates Inducible TNF Production in Humans," *Journal of Applied*

Physiology 103, no. 3 (September 1, 2007): 1007–11, https://doi.org/10.1152/japplphysiol.00147.2007.

226 **Some of the people who:** Peter Attia, *Outlive: The Science and Art of Longevity* (New York: Harmony, 2023).

11. Your Great Nerve: How to Talk to Your Doctor (the FAQs)

227 **epigraph:** Denise Mann, MS, "Cyndi Lauper Talks Vibrant Aging—And Staying a 'Working Girl' at 70," *The Healthy,* January 11, 2024, https://www.thehealthy.com/skin-health/news-cyndi-lauper-age-health-interview/.

Coda: The Clear and Present Future: Computer Chips, Not Medicines

247 **epigraph:** Dennis Gabor, Nobel Prize–winning physicist, *Inventing the Future* (New York: Alfred A. Knopf, 1964). Commonly misattributed to Abraham Lincoln: https://www.nprillinois.org/lincoln/2018-02-02/governor-goofs-honest-abe-quote.

247 **epigraph:** Simone de Beauvoir, *All Said and Done: The Autobiography of Simone de Beauvoir* (Boston: repr., Da Capo Press, 1994).

249 **"a life free of pain":** In this chapter, quotes and stories from Kelly Owens where I was not present are adapted from the unpublished writings of Kelly Owens, reprinted with her permission.

263 **Vagus Nerve Stimulation Using Implanted Devices:** Modified from: Yi-Ting Fang et al., "Neuroimmunomodulation of Vagus Nerve Stimulation and the Therapeutic Implications," *Frontiers in Aging Neuroscience* 15 (July 5, 2023), https://doi.org/10.3389/fnagi.2023.1173987.

Index

abdominal adiposity, 142
ABVN. *See* auricular branch of vagus nerve
acetylcholine (neurotransmitter)
  as difficult to measure, 198, 212
  parasympathetic nervous system and, 32, 33
  as "vagus stuff," 32, 76, 158, 198
action potentials
  compared to electricity, 74–75
  neurons and, 38
  releasing neurotransmitters, 76
  speed of, 75
acupuncture, 159–60
acupuncture maps, 160, 165–66
aerobic exercise, 41
AES. *See* auricular electrical stimulation
afferent nerve fibers, 51, 69, 146, 159
Alice in Wonderland syndrome, 84
Altmetric, 142–43
Alzheimer's disease, 239–40
AMA. *See* American Medical Association
American Heart Association, 185
American Medical Association (AMA), 139–40
Amsterdam University Medical Center, 4
amygdala, 133, 236
Andersson, Ulf
  headset-style taVNS of, 163, 172–74, 245
  HRV and RSA monitoring of, 171
  illness and recovery of, 155–58
Angelou, Maya, 155
animals, research using, 27–29
antidepressants, 128–29

anti-inflammatory drugs, 18, 64–65, 96–97, 100
anti-inflammatory healing reflex, 4, 66, 68–69, 112, 149
anti-seizure medications (ASMs), 85
anti-TNF monoclonal antibodies, 60, 62–65, 96, 112–13
anxiety
  inflammation and, 127
  sympathetic nervous system and, 36, 121, 122
  TENS and, 172
apps (mobile phone)
  for breathwork, 199–200
  for meditation, 187
Aristotle, 24, 26
Arnold, Friedrich, 159
Arnold nerve reflex, 159
arteriovenous malformation, 83
arthritis, cold therapy and, 214
artificial intelligence, 151, 243–44
ASMs. *See* anti-seizure medications
attention score, 143
Attia, Peter, 226
auricular branch of vagus nerve (ABVN), 159
auricular cough reflex, 159
auricular electrical stimulation (AES), 165
auricular VNS methods, 239
auriculomedicine, 160
auriculotherapy, 159–62
autoimmune diseases. *See also* Crohn's disease; rheumatoid arthritis
  inflammation and, 5, 93, 127

INDEX

autonomic nervous system
   function of, 12–13, 30
   Wim Hof Method influencing, 192, 195–97
axon, 38

baboons, 62
Bakken, Earl, 81
Bartholin, Caspar, 41–42
BAT. *See* brown adipose tissue
BCIs. *See* brain-computer interfaces
BDNF. *See* brain-derived neurotrophic factor
Beauvoir, Simone de, 247
behavior, 86, 125, 131–32
Beijing Normal University, 190, 198–99
Bell, William O., 86–87, 118
Bering-Breuer reflex, 53–54
Bernard, Claude, 43
Bevanda, Milenko, 104
bioelectronic medicine, 16–17, 74, 97, 113–14
biologics, 100, 113, 256–57
Blackburn, Elizabeth, 181–83, 186
blood pressure, 53, 122, 161, 171, 184
body mass index (BMI), 137, 139–40
body-mind integration method, 190
bradycardia, 79, 88
brain
   depression and, 119–20
   fMRI research on, 168, 183–84
   leaky gut and, 132–33
   vagus nerve and, 29, 225
   VNS affecting, 90–91, 186–87, 236–37
brain-computer interfaces (BCIs), 242
brain-derived neurotrophic factor (BDNF), 236–37
breathing
   breathwork and, 40, 190–203
   cold exposure and, 212–13
   vagus nerve fibers controlling, 56, 200–201
brown adipose tissue (BAT), 220
Bucholz, Richard, 102–5, 106
Bugut-Ante, 101, 102
burn injuries, 43–45
Butler, Robert, 220–21

Cajal, Ramón y, 150
Cakmak, Yusuf Ozgur, 165
cardiovascular disease, 33, 171–72, 184
cardiovascular system, 169
catecholamines, 122, 123, 216

CDAI. *See* Crohn's disease activity index
Cerami, Tony, 223
cervical plexus, 164
Chavan, Sangeeta, 42, 94. *See also* Feinstein Institutes
China Academy of Chinese Medical Sciences, 172
chromosomes, 186
chronic stress, 121, 127, 129–30, 132–33
clinical trials. *See also* randomized controlled trials; research
   approval process for, 95–96
   of breathwork for COVID-19, 200
   of cold therapies, 207
   of focused ultrasound, 144
   of meditation, 184–86
   size of, 208
   of SSRIs, 129
   of taVNS amd TENS, 163, 170, 172–73
   of VNS, 3–5, 17, 229
   of VNS for Crohn's disease, 111–12
   of VNS for depression, 116–19, 134–35
   of VNS for rheumatoid arthritis, 97–109
clinicaltrials.gov (website), 209, 229
cluster headaches, 171
CNI-1493 (anti-inflammatory molecule), 65–66
cold, 210–13, 217–18
cold therapies
   benefits of, 206–7, 219–20
   cytokines and, 213, 215–16
   dangers of, 207
   plunging, 206–7, 210–13
Columbia University, 181, 224
compliance, 139, 152–53
Cornell University Medical Center, 59, 62
Corning, J. L., "neck truss" of, 77–78
cortisol, 196
cost of VNS, 112–13, 235
COVID-19, 200
C-reactive protein (CRP), 124
Critical Therapeutics, 95
Crohn, Burrill, 109–10
Crohn's disease activity index (CDAI), 111, 112
Crohn's disease
   symptoms of, 109–10, 250–51
   treatments for, 110, 251
   VNS and, 109–13, 170
CRP. *See* C-reactive protein
Cyberonics, 83, 88
cyclic hyperventilation, 199
cymba concha, 158, 159, 164, 166

296

INDEX

cytokines
  biologics and, 100
  chronic stress and, 123
  cold therapy and, 213, 215–16
  cytokine storms and, 125–26, 148, 166
  Hof and lowering levels of, 194
  immune system and, 60, 196–97
  leaky gut and, 132
  suppression using pFUS, 149–50
  suppression using taVNS, 174–75
  suppression using TENS, 166–68

Dalai Lama, 181–83
data privacy, 243
death. *See* mortality rates
depression
  brain and, 119
  cytokine-induced, 126, 130
  experience and symptoms of, 115–16, 119–21
  frequency of, 115
  risk factors for, 128
  SERT gene and, 129–30
  sickness syndrome and, 126
  TENS and, 172
  theories of, 127, 128–30, 133–34
  VNS and, 116–19, 230–31
diabetes, 140, 223
diaphragmatic breathing, 40–41, 190–91
Dickens, Charles, 3
disease
  as impaired homeostasis, 56
  inflammation and, 6, 93, 107
disease-modifying antirheumatic drugs (DMARDs), 99–100, 102
disequilibrium, 56
diving reflex, 55, 211
DMARDs. *See* disease-modifying antirheumatic drugs
Dragoje, Pero (patient), 102–5
Dravet syndrome, 89
Duchenne muscular dystrophy, 173–74

ear, 157–58, 164
E. coli (bacterium), 45, 63
EEG. *See* electroencephalograms
electricity
  action potentials and, 38
  dangers of, 73–74
  heart and, 31–32
  nervous system and, 31, 66
electroencephalograms (EEG), 78, 85
electrons, neurons and, 74–75

endotoxins
  experiments using, 125–26
  in testing cytokine production, 223–24
  in testing TNF production, 195–96, 216–17
  in testing Wim Hof Method, 195–97
epilepsy
  Hippocrates on, 83
  symptoms of, 84–85
  therapies for, 85–86
  VNS and, 82–91, 93–94, 230
Epilepsy Association of North Carolina, 87
epinephrine, 79, 122, 195, 196, 210
ethics, 242–43
exercise, 36, 40–41, 221–26

far-field evoked brain stem potentials, 168
FDA. *See* Food and Drug Administration
Feinstein Institutes
  animal research at, 27–29
  holistic approach of, 94
  mission of, 17
  research on CNI-1493 at, 65
  research on Hof at, 192–94
  research on inflammation at, 93–95, 149, 216–17, 223–24
  research on pFUS probe at, 149
  research on RA using taVNS at, 174–75
  research on RA using VNS at, 97–109
  research using focused ultrasound at, 141–42, 165–67
  staff members of, 42, 94, 254
fight-or-flight response
  cold exposure and, 210–12, 218
  inflammation suppression and, 194–98
  norepinephrine and, 32–33
  rest-and-digest response and, 34–36
  sympathetic nervous system, 13, 122–23
fMRI. *See* functional MRI
focused ultrasound. *See also* peripheral focused ultrasound
  inflammation and, 147–50, 165–67
  operation of, 143–44
  research using, 141–42, 165–67
  vagus nerve and, 141, 238–39, 244
  vagus nerve mapping and, 150–51
  weight loss and, 141–42, 240
Food and Drug Administration (FDA)
  approval of TENS in anxiety, depression by, 172
  approval of TENS for headaches by, 171
  approval of TENS in opioid withdrawal by, 170–71

INDEX

Food and Drug Administration (*cont.*)
   approval of VNS for depression by, 134–35, 229
   approval of VNS for epilepsy by, 88–89, 229
   approval process of, 95–96
   breakthrough designation of, 5, 107, 237
   classification of TENS by, 163
Fournie, Nick (patient), 115–19, 135
Frankenstein (film), 81
Frankenstein (Shelley), 73
functional MRI (fMRI), 168
future
   of bioelectronic devices, 151
   planning, 120–21

GABA. *See* gamma-aminobutyric acid
Gabron, Dennis, 247
Galenus, Claudius (Galen), 24–26, 41, 57, 210
Galvani, Luigi, 21
gamma-aminobutyric acid (GABA), 90, 168–69, 236
gastrointestinal system, 139
gate control theory, 162
GE Global Research, 141, 148, 149
GE HealthCare, 147
German, William J., 9, 10
glioblastoma multiforme, 7–8
GLP-1. *See* glucagon-like peptide 1
GLP-1 agonists (medications), 140–41, 146
GLP-1 receptors, 146
glucagon-like peptide 1 (GLP-1), 140, 239, 240
glucocorticoids, 122
gluconeogenesis, 122
glutamate, 90, 236
Good, Robert A., 68
Great Expectations (Dickens), 3
great nerve. *See* vagus nerve
gut-brain axis, 238, 239
gut microbiome, 132, 145, 146, 239

Hamilton Depression Rating Scale, 172
Harris, Sam, 187
Harvard Medical School, 39–40, 56, 161, 172, 190, 198–99
HeadSpace (meditation app), 187
healing
   anti-inflammatory healing reflex and, 4, 66, 68–69, 112, 149
   process of, 52–55, 58
health consciousness, 180

heart, 31–33
heart rate variability (HRV)
   deep breathing and, 213
   exercise and, 221–22
   sympathetic nervous system and, 36, 122
   TENS and, 171
hemoglobin A1C (blood test), 223
hepatic glucose sensing reflex, 54–55
hippocampus, 236
Hippocrates, 24, 83, 137, 209
HMGB1 (inflammatory protein), 158
Hof, Wim, 191–95, 206, 209
homeostasis
   as balance of systems, 36, 50–52, 55
   vagus nerve and, 55–56, 138
HRV. *See* heart rate variability
hypertension, 161, 199–200
hyperventilation, 212
hypoventilation, 212
hypoxia, 195

IBD. *See* inflammatory bowel disease
IL-1 (cytokines), 123, 126, 131–32
IL-6 (cytokines), 123, 126, 200
IL-10 (cytokines), 195
immune system
   cells in, 58–59
   inflammation and, 57–58
   sensory systems and, 68–69
   suppression of TNF release by, 94
   TENS and, 169
   vagus nerve and, 15
   Wim Hof Method and, 192, 195–97
immunosuppressive drugs, 6, 58, 100
infections, bacterial, 45, 156
inflammation
   cold and, 213–20
   diseases of, 6, 93, 107
   excessive, 59–65
   focused ultrasound and, 147–50
   following surgery, 156
   functions of, 57–58
   leaky gut and, 132–33
   meditation and, 185–86
   rebalancing, 93–104
   SSRIs and, 129
   stress, anxiety, and, 123
   TENS and, 170
   testing for, 223
   VNS and, 90
inflammatory bowel disease (IBD), 96, 238
Institutional Review Board, 192–93
interferons (biological drugs), 126

INDEX

interleukins (biological drugs), 126
International Journal of Hyperthermia, 215
ion channels, 74–75
Istituto di Fisiologia, 78

Jamison, Kay Redfield, 115
Janice (patient), 43–45, 61

Karolinska Institute, 155
Kincaid, Toney (patient), 83, 86–88
knee-jerk reflex, 47–48

Landau-Kleffner syndrome, 89
Largus, Scribonius, 162
larynx, vagus nerve and, 26
lasers, 39
Lauper, Cyndi, 227
leaky gut syndrome, 132–33
Lennox-Gastaut syndrome, 89
Levine, Yaakov, 97, 102–3
Lillehei, C. Walton, 81
lipolysis, 122
lipopolysaccharide (LPS), 45, 132, 148
liver, VNS and, 240
Loewi, Otto, 31–32, 66, 76
Lowry, Stephen, 125–26
LPS. *See* lipopolysaccharide
lymphocytes, 58–59, 99, 148

macrophages, 61, 99, 148
Margie (patient), 109–13
MBIs. *See* mindfulness-based interventions
McNally, William, 11
medications
 bioelectric therapies replacing, 3–19
 in conjunction with VNS, 234–35
 for obesity, 140
 for rheumatoid arthritis, 99–100
 side effects of, 251–52
meditation
 apps for, 187
 benefits of, 186–87, 189
 methods and types of, 183, 187–88, 193
 research into, 180–81, 183–86
Mediterranean Society of Acupuncture, 160–61
Medtronic, 81, 82
Melzack, Ronald, 162
membrane potential, 74–75
Memorial Sloan Kettering, 59, 60
methotrexate, 100, 102, 106
mice
 obesity studies and, 138–39, 141–42
 in research, 27–29
microbiome-gut-brain axis, 146
migraine headaches, 171
mind-body connection, 67
mindfulness-based interventions (MBIs), 185–86
monoclonal anti-TNF antibodies, 100, 256–57
monocytes, 99, 148
mood, 118–19, 225
mortality rates, 6, 33–34, 222–23

National Institute on Aging (NIA), 220–21
National Institutes of Health, 150
Nature (journal), 16, 64–65, 96
"neck truss," 77–78
nervous system. *See also* parasympathetic nervous system; sympathetic nervous system; vagus nerve
 autonomic, 12–13
 electricity and, 31
 history of research into, 24–26
 sensory nerves in, 12, 46–47
nervus vagus (Lat. wandering nerve). *See* vagus nerve
neural regeneration, 239–40
neurons, 39, 74–75
neuroscience, 150
neurotransmitters, 32–33, 38. *See also* individual names of neurotransmitters
neutrophils, 61
New York Hospital, 43
nociceptors (sensory neurons), 210, 219
Nogier, Paul, 160
noninvasive VNS methods, 152–53, 240–41
norepinephrine
 sympathetic nervous system and, 32–33, 122, 196, 197
 brown adipose tissue and, 220
North Shore University Hospital, 66–67
Northwell Health, 149
Novo Nordisk, 147
NTS. *See* nucleus tractus solitarius (NTS)
nucleus tractus solitarius (NTS), 159

obesity, 139–40, 145–46
"On the Sacred Disease" (Hippocrates), 83
opioids, 90, 170–71
opsins (light-sensitive proteins), 39
optogenetics, 39–41
Owens, Kelly (patient), 3–5, 17, 249–55

INDEX

pacemakers, 79–82
pain
  cold therapy and, 206–7, 210, 214
  gate control theory and, 162
  as inflammatory response, 58
  TENS and, 162–63
  VNS and chronic, 237
pancreatic tumor, 155–56
parasympathetic nervous system
  breathwork and, 40–41, 190
  cold response in, 211–12
  high vagal tone and, 36
  rest-and-digest response and, 13, 33, 122
  sympathetic nervous system and, 34–37, 197–98
Parkinson's disease, 239–40
Pascal-Wager strategy, 203
Pasteur, Louis, 98, 106
patellar reflex, 47–48
pediatric nephrotic syndrome, 170
Peijing Rong, 172
peripheral focused ultrasound (pFUS). *See also* focused ultrasound
  inducing anti-inflammatory reflex, 148–49
  VNS and, 144–47
pFUS. *See* peripheral focused ultrasound
phagocytes, 58–59, 61
pharmaceutical industry, VNS and, 18, 180, 255
physicians, VNS and, 245
Pickkers, Peter, 195
piezoelectric crystals, 143
placebo effect, 254–55
planning, depression and, 120–21
polio vaccine, 98, 106
prednisone, 102, 110, 251–52
pre-existing conditions, VNS and, 230
Proceedings of the National Academy of Sciences, 107, 195
psoriasis, 227
Puddicombe, Andy, 187
Purdue University, 224

RA. *See* rheumatoid arthritis
rabies vaccine, 98, 106
Radboud University Medical Center, 195
randomized controlled trials (RCTs), 207–9. *See also* clinical trials
RCTs. *See* randomized controlled trials
reciprocal inhibition, 48
reflexes
  anti-inflammatory healing reflex, 4, 66, 68–69, 112, 148–49
  Arnold nerve reflex, 159
  auricular cough reflex, 159
  baroreceptor reflex, 53
  Bering-Breuer reflex, 53–54
  coordination between, 48
  coughing, 50
  diving reflex, 55, 211
  healing, 15
  hepatic glucose sensing reflex, 54–55
  inflammation as, 57
  knee-jerk, 47–48
  patellar, 47–48
  reciprocal inhibition and, 48
  respiratory sinus arrhythmia, 54
  sensory feedback and, 48–49
  vaso-vagal reflex, 54
research. *See also* clinical trials
  animal use in, 27–29
  on benefits of exercise, 222–25
  challenges of vagus nerve in, 169–70
  on cold in arthritis treatment, 214
  on cold exposure, 215–18
  on cytokine storms, 125–26, 148–49
  on diaphramatic breathing, 190–91, 198–99
  on electricity and nerves, 30–33
  on meditation, 180–81
  on spleen nerves using pFUS, 149–50
  on TNF using baboons, 62–65
  using taVNS and TENS, 161, 165–68, 170
  on vagus nerve in sheep, 36–37, 197–98
  on VNS and epilepsy, 88–89
  on VNS in obese mice, 138–39
  on Wim Hof Method, 195–97
RESET-RA study, 107–8
respiratory alkalosis, 195
respiratory sinus arrhythmia (RSA), 54, 171–72, 213
rest-and-digest response
  acetylcholine and, 32–33, 158
  fight-or-flight response and, 34–36
  parasympathetic nervous system and, 13, 37, 122
rheumatoid arthritis (RA)
  taVNS, TENS, and, 170, 174–75
  treatments for, 99–100, 108–9
  VNS and, 97–109, 231, 237
RSA. *See* respiratory sinus arrhythmia

Salk, Jonas, 98, 106
Sangeeta. *See* Chavan, Sangeeta

INDEX

satiety signals, 145
Scientific Reports, 142
selective serotonin reuptake inhibitors (SSRIs), 128–30
sensory nerves, 46–47, 68–69
serotonin, 128–29
SERT gene, 129–30
SetPoint Medical
    clinical trials of, 96–97, 100–109
    founding of, 95
    RESET-RA study of, 107–8
sheep, 36–37, 197–98
Shelley, Mary Wollstonecraft, 73
Sherrington, Charles, 21
shock, TNF and, 64
sickness syndrome, 125
Sloan, Richard, 224
sodium channels, 38
sonogram, 144
spleen, 148–50
Spock, 179
SSRIs. *See* selective serotonin reuptake inhibitors
steroids, 110, 173
Stokes-Adams disease, 79–80
stress. *See* chronic stress
Suzuki, Shunryu, 180
sympathetic nervous system
    cold response in, 211–12
    "fight-or-flight" response and, 13
    norepinephrine and, 32–33, 122, 196–97
    parasympathetic nervous system and, 34–37, 197–98
    stress, anxiety, and, 121, 123

taVNS. *See* transcutaneous auricular vagus nerve stimulation
T-cells, 148–49
telomeres, 186
TENS. *See* transcutaneous electrical nerve stimulation
Tenzin Gyatso, the XIVth Dalai Lama, 181–83
thalamus, 236
Thomas, Lewis, 57–58, 93
T-lymphocytes, 148–49
TNF. *See* tumor necrosis factor
trained immunity, 58–59, 123
transcutaneous auricular vagus nerve stimulation (taVNS)
    acupuncture and, 159–60
    Andersson research on, 157–58, 163, 172–74, 245

    clinical trials of, 163
    depression and, 172
    hypertension and, 161
transcutaneous electrical nerve stimulation (TENS)
    Andersson research using, 157–58
    FDA classification of, 163
    GABA and, 168–69
    imprecision of, 164–65
    over-the-counter, 153–54, 240–41
    parameters for, 174
    research into, 165–69
    rheumatoid arthritis and, 174–75
    sensation of, 166
    theory of, 162–63
transcutaneous VNS methods, 238–39
transducer, 143, 149
transient receptor potential (TRP), 210
TRP. *See* transient receptor potential
tumor necrosis factor (TNF)
    acetylcholine and, 158
    in cytokine storm, 126
    exercise and, 224–25
    as inflammatory cytokine, 60–61, 123, 132
    suppression of, 65–66, 94, 106–7, 148–50

University Clinical Hospital in Mostar, 101
University of Gothenburg, 88
University of Otago, 165
University of Washington, 222

vaccines, 98
vagal tone
    breathwork and, 190, 200
    exercise and, 222
    high, 36
    increasing, 22, 40, 53, 171, 184–85, 190
    as integrative function, 34
    low, 36, 171
    measurement of, 36, 184
    meditation and, 184–85
vagotomy, 146–47
Vagus ECG Smartwatch, 171
Vagus Health, 171
vagus nerve. *See also* nervous system
    anatomy of, 13–14, 25–26, 29
    body weight and, 138–39
    breathing exercise to stimulate, 40–41
    challenges in researching, 169–70
    disease response and, 131–32
    ear signals to, 159
    exercise and, 36, 220–26

INDEX

vagus nerve (cont.)
  functions of, 22, 50
  hypotheses about, 200–201
  immune system and, 15
  meditation and, 181–83
  names for, 13–14, 41–42
  parasympathetic nervous system and, 13
  physiology of, 30–33, 38
  roles in depression of, 133
  satiety signals and, 145
  sensory nerve signals in, 46–47, 51, 69
vagus nerve stimulation (VNS). *See also*
    focused ultrasound; transcutaneous auricular vagus nerve stimulation (taVNS); transcutaneous electrical nerve stimulation (TENS)
  artificial intelligence and, 243–44
  auricular or ear-based, 239
  blocking TNF production, 66–67
  brain regions and, 236–37
  chronic pain and, 237
  compliance and, 139
  cost of, 112–13, 235
  Crohn's disease and, 111–13
  definition of, 228–29
  depression and, 134–35
  early practice of, 77–78
  effectiveness of, 230–31
  electrical pulse generator and, 75–76
  epilepsy and, 66, 78–79, 82–91, 93–94
  FDA approval of, 88–89, 134–35, 229
  future of, 255–58
  IBD and, 238
  medications and, 234–35
  mood, behavior, and ethics of, 242–43
  neural regeneration and, 239–40
  noninvasive methods of, 139, 144, 238–39, 244
  pacemaker and, 79, 82–83
  physiological effects of, 76
  pre-existing conditions and, 230
  procedure for implanting, 89, 231–32
  quality of life and, 235
  as removable, 233–34
  rheumatoid arthritis and, 98–99, 101–9, 237
  side effects of, 103, 106, 232, 234
  theories of effectiveness of, 90
  therapy using, 232–33, 236
  TNF production and, 94–95, 106–7
  types of, 164
  weight loss and, 240
vagusstoff (Ger. vagus stuff). *See* acetylcholine
"vagus stuff." *See* acetylcholine
vaso-vagal reflex, 54
VNS. *See* vagus nerve stimulation
VO2 maximum testing, 224–25
vocal cords, 26

Waking Up (meditation app), 187
Wall, Patrick, 162
wandering nerve. *See* vagus nerve
Warren, H. Shaw, 95, 97
Washington University School of Medicine, 116
Watkins, Linda, 131–32
Weber, Eduard and Ernst, 30–31, 66
weight loss
  focused ultrasound and, 138–39, 141–42, 145
  medications for, 140
Western diet, 141–42
Whipple procedure, 156
white blood cells
  immune system and, 99
  inflammation and, 58–59, 61, 94
  research using, 216–17
  spleen and, 148
  trained immunity and, 58–59, 123
whole blood monocyte culture assay, 149
Wim Hof Breathing Method, 191–98, 202–3
World Health Organization (WHO), 139

Zabara, Jacob, 82
Zanchetti, Alberto, 78–79, 82
Zanos, Stavros, 150, 151
Zen Mind, Beginner's Mind (Suzuki), 180
Zitnick, Ralph, 100–101
Zoll, Paul, 79–80